Science and Invention

Science and Invention

Ray Spangenburg and Diane K. Moser

Facts On File, Inc.

American Historic Places: *Science and Invention*

Copyright © 1997 by Ray Spangenburg and Diane K. Moser

Facts On File, Inc.
11 Penn Plaza
New York NY 10001

Library of Congress Cataloging-in-Publication Data

Spangenburg, Ray, 1939–
American historic places : science and invention / Ray Spangenburg and Diane K. Moser
p. cm.
Includes bibliographical references and index.
ISBN 0-8160-3402-8
1. Discoveries in science—United States—History. 2. Inventions—United States—History.
3. Historic places—United States.
I. Moser, Diane, 1944– . II. Title.
Q180.55.D57S63 1997
609.73—dc21   97-9873

You can find Facts On File on the World Wide Web at http://www.factsonfile.com

Text design by Cathy Rincon
Cover design by Dorothy Wachtenheim
Layout by Robert Yaffe
Illustrations on pages vi, 6, 17, 29, 42, 48, 69, 95, 111 by Jeremy Eagle

This book is printed on acid-free paper.

Printed in the United States of America

RRD FOF 10 9 8 7 6 5 4 3 2 1

To the memory of

Joseph Henry

and the true spirit

of early American science

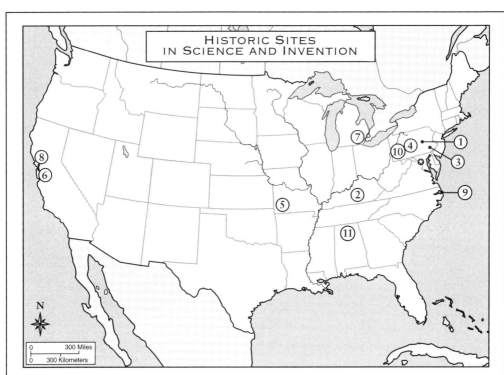

HISTORIC SITES
IN SCIENCE AND INVENTION

N

| 0 | 300 Miles |
| 0 | 300 Kilometers |

1 Joseph Priestley House,
   Northumberland, Pennsylvania

2 McDowell House and Apothecary Shop,
   Danville, Kentucky

3 Hopewell Furnace National Historic Site,
   Elverson, Pennsylvania

4 Allegheny Portage Railroad National Historic
   Site, Cresson, Pennsylvania

5 George Washington Carver National Monument,
   Diamond, Missouri

6 Lick Observatory, San Jose, California

7 Thomas Edison's Menlo Park Laboratory,
   Dearborn, Michigan

8 Luther Burbank Home and Gardens,
   Santa Rosa, California

9 Wright Brothers National Memorial, Kill
   Devil Hills near Kitty Hawk, North Carolina

10 Rachel Carson Homestead,
   Springdale, Pennsylvania

11 U.S. Space and Rocket Center,
   Huntsville, Alabama

CONTENTS

PREFACE TO THE SERIES

History doesn't have to be dry or stuffy. And it isn't exclusively about military skirmishes and legislative proclamations—they make up only a small part of it. History is the story of life events that happened to people who cared as passionately about their lives as we care about ours. And it's the story of events that often continue to shape and influence our lives today. But getting to the human side of these stories isn't always easy. That's why there's nothing like visiting the place where an event actually occurred to get the feel of what it all meant.

The study of historic places—what happened at a particular site and how the lives of the people there were affected—has emerged as a great way to approach history, to "relive" the experience and open up to the immense diversity of American culture. Every community and region is rich in such places—places that highlight real stories about real people and events. Even if you can't actually visit such a place, the next best thing is to go there through pictures and words. Use this book and the other books in this series as jumping off points and look around your community for places where you can experience the world of the people who once lived in your own region—and begin exploring!

♦ ♦ ♦

Each volume in this series explores a different aspect of U.S. history by focusing on a few select places. This book takes a look at historic places in the United States associated with science and invention. Of course, choosing exactly which places to focus on in each book was one of the most difficult tasks of this project. We limited our choices to sites that had either been restored or maintained in authentic historic condition—many are National Historic Landmarks, chosen by the U.S. government to be preserved for their historic significance. We also tried to include examples from a wide variety of locations, events and experiences, types of sites and time periods. We then limited our selections to just a few. But many other fascinating places exist throughout the country, and that's why we mentioned other related sites at the end of some chapters (under Exploring Further) and added a list of additional sites at the back of the book (More Places to Visit).

Each chapter begins with information about the site At a Glance. Then we explore the place—what it's like and who lived there, how the place related to that person's life and work, and what it's like to visit there today. We also look closely at one feature of the site in "A Close-Up" section, followed by a section recapping how the site came to be a protected historic site (Preserving It for the Future). A list of books and other resources concludes each chapter (Exploring Further), directing readers to either a broader or closer view of the development of science and technology in the United States.

Exploring historic sites not only provides a way to experience past events with fresh vividness and immediacy, it also offers a way of seeing the past through new eyes, through the eyes of those who lived it. For this adventure—and it can prove to be a lifetime adventure—this series will have accomplished its purpose if it provides the springboard for future explorations. In the words of an old Gaelic greeting, "May the wind be always at your back and may the road rise up to meet you," as you travel down these avenues of historical experience.

ACKNOWLEDGMENTS

Warm thanks to all who have helped us make this book better, to name just a few: Alberta Moynahan at McDowell House, Reid Miller at Allegheny Portage Railroad, Mark Tomlinson at the Rachel Carson Homestead, Lana Henry and Tammy Benson at the George Washington Carver National Monument, Robert Eaton of the Pennsylvania Historical and Museum Commission, Cathy Stevenson at the Luther Burbank House and Gardens, Darryl Collins at the Wright Brothers National Memorial, and Jeffrey Collins at the Hopewell Furnace National Historic Site. At Facts On File, special thanks to our editor, Nicole Bowen, for her steady professionalism and incomparable commitment to excellence; to copy editor Michael G. Laraque; and to our former editor, James Warren, who helped us conceptualize the series.

INTRODUCTION

When early European settlers journeyed across the Atlantic to the shores of North America, they brought with them, in their minds and in their notebooks, much of the science and technology that they used in those early days. Iron making, for example, as developed in ironworks like the Hopewell Furnace in Pennsylvania discussed in this book, was already prevalent in Great Britain and Europe, and the ironworks in North America were begun to meet colonists' needs by people who had learned the trade before arriving in the New World.

The lack of roads isolated the early American colonial communities so that scientific ideas and inventions were not traded and shared very much. And most people had to spend all their time trying to survive in the wilderness, reserving their creative energies for making the tools they needed for their own work. As a result, American innovation really didn't begin to play an important part until nearly 150 years after the first European settlers arrived.

Nonetheless, these were courageous pioneers, and when necessity required, as it did for Ephraim McDowell in the Kentucky wilderness, they called upon all the knowledge, skill, and talent they had available. In

McDowell's case, these were in plentiful supply, as shown by the story of his unprecedented surgery and the fortitude and determination of his patient, Jane Todd Crawford. A visit to the house in Danville, Kentucky, where the surgery took place serves as a reminder of how different were their lives from ours—and how far medical science has come.

By the late 18th century, roads had begun to form a network among the colonies, and communication with Europe—at least for philosopher-scientists such as Benjamin Franklin and Thomas Jefferson—made it possible for American scientists to participate actively in international scientific development. Jefferson was deeply interested in scientific research, studied agriculture, and experimented with new varieties of grain. He also studied and classified fossils unearthed in New York State at a time when this branch of science was in its infancy. Franklin founded America's first scientific society, the American Philosophic Society, in 1743, and inventions such as the lightning rod and the Franklin stove, as well as his work with electricity, are well known. Both Jefferson and Franklin knew English chemist Joseph Priestley and encouraged him to make his home in the United States. In 1795 Priestley arrived to settle in the little town of Northumberland, Pennsylvania, where the house he built still stands.

Although Joseph Henry (1797–1878) is less well known, he took up the tradition of Franklin and Jefferson for the following generation and formed the foundation for the growth of scientific development in the United States in the 20th century. Henry, whose former home has become part of the Princeton University campus in New Jersey, embodies the true spirit of early American science: an "enlightened" mind in the best sense—pursuing science for the sake of knowledge and not for profit. He took out no patents and missed out on credit for several discoveries because he wouldn't rush his results into print. He invented a powerful electromagnet and an electric motor, which set the stage for the work of Thomas Edison, whose invention of a practical incandescent bulb in 1879 launched the modern era of electricity. Henry encouraged young scientists and fostered open communication about scientific work, founding several magazines and societies toward this end, including the Smithsonian Institution, which has become a center of education and research in all fields of science, technology, and history.

By the 19th century, American inventors began to come into their own as well. The very richness of the North American continent had in some ways worked against the development of American innovation at first. But by the 1820s and 1830s, engineering began to become important as regions sought to build connecting canals and railroads, such as the Pennsylvania Mainline of Public Works, of which the Allegheny Portage Railroad was part—now preserved as a National Historic Site.

The new nation's use of slave labor was one of its gravest mistakes morally, philosophically, and politically. The practice had the additional negative effect of slowing the wheels of technological development—because slave labor was cheap, particularly in agricultural areas, little need was felt for labor-saving devices. After the Civil War ended in 1865, emancipation and Reconstruction began to cause changes in that situation. And one man, agronomist George Washington Carver, did much to reform the poor agricultural practices still prevalent in the South as late as the 1890s. His work is commemorated at his birthplace in Diamond, Missouri, as well as at the Tuskegee Institute in Alabama.

In the meantime, successful entrepreneurs like James Lick of San Francisco began to think in terms of endowing great scientific facilities. Lick generously bequeathed his fortune to build the first mountaintop observatory, high on Mount Hamilton in California. Lick Observatory remains one of the world's great astronomical observatories, and installation of its first telescope provided the rare opportunity for a team of American lens makers, Alvan Clark and his son Alvan Graham Clark, to emerge, as well, in a field that had long been dominated by British technologists.

The period from the 1840s to the turn of the century bristled with great technological breakthroughs—Samuel Morse's telegraph, Alexander Graham Bell's telephone, the Remington typewriter, and many more. An inordinate number were made by one man: Thomas Alva Edison, whose early laboratory, once located in Menlo Park, New Jersey, has been meticulously reconstructed at Greenfield Village at the Ford Museum in Dearborn, Michigan. His later laboratory, at West Orange, New Jersey, as well as his home there are also open to the public as a National Historic Site. Visitors to Fort Myers, Florida can see his winter home, where he also had a laboratory (this was a man who never stopped working), and his birthplace in Milan, Ohio is open to visitors as well.

In the biological sciences, Luther Burbank combined the scientist's powers of observation and often intuitive understanding of observed phenomena with the inventor's creativity and dynamo-like perseverance. Visitors can appreciate some of the results of his tireless adventures with plant hybrids at his home in Santa Rosa, California.

Perhaps no single technological event symbolizes the opening of the 20th century better, however, than the Wright brothers' first flight at Kitty Hawk, North Carolina. This 1903 triumph transformed transportation worldwide and in one exhilarating moment ushered in the age of air travel, following close on the miracle of telephone communication to shrink the vast distances between cities, states, and countries.

Zoologist Rachel Carson's calm, cautionary voice and her rhapsodic love of nature remind us of both the appreciation of our planet and the concern with which we have learned to regard our scientific triumphs in the 20th century. Both are brought to mind by the unassuming little house where she grew up in Springdale, Pennsylvania.

Finally, the U.S. Space and Rocket Center in Huntsville, Alabama commemorates the work of Wernher von Braun and the extraordinary historical events leading up to the launching of the nation's first satellites, spacecraft, and space travel, including the Apollo missions to the moon, space shuttle flights, and the international space station.

Vivid curiosity is the mainstay of science, and determined application of principles is the mainstay of technology. These are stories of courage—McDowell, the Wright brothers, Carson, and von Braun. Of enterprise—Hopewell and the Allegheny Portage Railroad. And of curiosity, determination, and perseverance—Priestley, Carver, the Lick Observatory, Edison, and Burbank.

From these stories we can gain not only greater appreciation for the present but, most important, a key to forming a better future. The great strides in progress made in the fields of science and technology in the last 250 years have bestowed on us not only enormous gifts but, as Rachel Carson pointed out so powerfully, great responsibilities as well. To control the tools created by modern science and technology with intelligence, we must be concerned not only with facts but also with values and purpose. And we can begin by seeing the world through the eyes of those who have created the science and technology upon which we have built our society and culture —through the historical places in which they lived and worked.

For more places and sources to explore, we encourage you to look at the Exploring Further section at the end of each chapter and More Places to Visit and More Reading Sources at the end of this book.

Joseph Priestley House

AMERICAN HOME AND LABORATORY OF CHEMIST JOSEPH PRIESTLEY
Northumberland, Pennsylvania

AT A GLANCE

Built: 1794–98

Home of chemist Joseph Priestley, 1794–1804

This is the place to which Joseph Priestley fled after persecutors in his native England burned his house, laboratory, and books in 1791. Priestley and his wife planned this stately frame house near the Susquehanna River, setting aside one wing to serve as a laboratory for Priestley, whose body of scientific work included the identification of oxygen in 1774.

Address:
Joseph Priestley House
472 Priestley Avenue
Northumberland, PA 17857
(717) 473-9474

Joseph Priestley is acknowledged as the discoverer of oxygen, but he was not entirely honored in his own land. His books and home in England were destroyed by those who opposed his political and religious ideas, and he accepted the urging of his scientific friends and family in the United States to make a new home in Pennsylvania. There, he spent the last 10 years of his life, continuing with his scientific work, as well as his political and religious writing.

I do not think there can be, in any part of the world,
a more delightful situation than this.

—Joseph Priestley,
describing Northumberland, Pennsylvania

♦ ♦ ♦ ♦ ♦

Like many others, Joseph Priestley (1733–1804) came to American shores reluctantly, fleeing oppression, and hoping for greater freedom and a better life. In his native England he had become noted as a scientist and Unitarian theologian. (Unitarianism is a Christian religion that focuses on a single Divine Being instead of accepting the doctrine of the Trinity.) But he had also attracted criticism for his support of the early stages of the French Revolution, his criticism of the English government, his disagreements with the Church of England, and his opposition to the practice of slavery. In 1791, in an incident known as the Riot of Birmingham, a mob of reactionaries had burned his house, his laboratory, and his books.

After three years of attempting to rebuild his life in England, tired of controversy and dissension, and ready for peace and solace, Joseph Priestley and his wife Mary had packed up what was left of his beloved books and laboratory equipment and journeyed by ship across the wide Atlantic Ocean, fleeing England. They sought a political climate where Priestley would be free to practice his religion and to speak his mind.

Priestley gave one last sermon in England before leaving, in which he stated, "I do not pretend to leave this country, where I have lived so long and so happily, without regret; but I consider it as necessary and I hope the same good

Joseph Priestley's house in Northumberland ca. 1980 (Photograph courtesy of the Pennsylvania Historical and Museum Commission, 1996)

providence that has attended me hitherto will attend me still. When the time for reflection shall come, my countrymen, I am confident, will do me justice."

Joseph and Mary Priestley sailed on April 8, 1794 and, after a long and difficult passage, arrived in New York City on June 4. They were greeted by New York's governor, George Clinton, as well as members of a number of learned societies. Two weeks later, they traveled to Philadelphia, which was the nation's capital and largest city at the time. There, they were encouraged to settle; the many advantages, from religious to scientific, made Philadelphia the natural choice.

Priestley's sons, Joseph Jr., William, and Henry, who had already settled in America, were planning a Unitarian colony near the tiny rural town of Northumberland, Pennsylvania. So Priestley and his wife journeyed there, apparently with the idea that frequent trips to Philadelphia—much like trips that Priestley had often made in a few hours to London—could keep him in touch with his scientific and religious contacts. But as they traveled for five days on poor roads, over mountains and across rivers into the near-wilderness of central Pennsylvania, they realized that return trips to Philadelphia would be rare. One tavern and inn they stopped at was so unpleasant, according to Priestley's description, that he and Mary decided they would be more comfortable sleeping in the wagon.

Finally, they arrived in Northumberland, where they were enchanted by the beauty and tranquillity of the countryside. Plans for the Unitarian community fell through, but the Priestleys decided to settle in Northumberland anyway, where they could be near their sons and where Joseph could spend his time quietly writing. So in the fall of 1794 they bought land between two branches of the Susquehanna River and soon began building their home in America.

Inside, Priestley's house conveyed a simple dignity and warmth, as shown in this view of the dining room, ca. 1982. (Photograph courtesy of the Pennsylvania Historical and Museum Commission, 1996)

Mary Priestley, in particular, took great delight in the planning of the structure, which included an arched entranceway and displayed the elegant balance and symmetry of the Federal-style architecture of the time. Construction began in 1795, using local workers as well as carpenters imported from Philadelphia. Great care was taken, with each piece of timber kiln-dried at the site before it was used. Unfortunately, Mary died in 1796, a year before the house was completed, and never lived in her new home.

By 18th-century standards Priestley's house became an opulent addition to the sleepy little mid-Pennsylvania town that boasted only 100 residents at the time. The house still stands stately yet modest, its two-story facade and balustraded roof deck reflecting the simple, straight lines of the style of the period. Even today, coming into the main entrance, with its semicircular fanlight above, one can still easily imagine Priestley warmly welcoming guests in the parlor—although few visitors arrived from Philadelphia, very likely one or more of Priestley's parishioners would be visiting. He often held services in his home and, through his sermons and copious writing, he did much to further the cause of Unitarianism in the new nation. Most of his 150 books, one of which was publicly burned in England in 1785, dealt with religious and educational subjects.

In the library of his house, visitors can view Priestley's massive globe, which reflects his interest in geography and education. While teaching as a professor in England in his twenties, he had introduced modern history, constitutional law, economics, and political science into the curriculum well before they were widely taught, and he encouraged his students to question what they were taught and think for themselves. He was a man of the Enlightenment, one of the liveliest periods of Western intellectual history, characterized by a belief that the use of reason could overcome many of humanity's obstacles. And his "curriculum for a Civil and Active Life" was inspired by his commitment to reason and knowledge as guiding principles of education. Thomas Jefferson, who shared these ideals with Priestley, sought his advice when he founded the University of Virginia.

It's easy to sense the reflective mind of Priestley in this room where he often read, wrote, or played a quiet game of chess with his son Joseph, who also lived in the house with his wife Elizabeth and their children. The chessboard still remains to remind visitors of the contemplative life Priestley

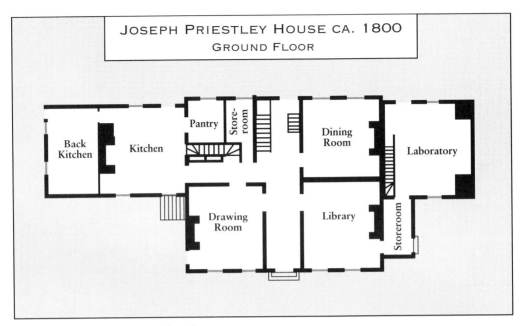

Back Kitchen

Kitchen

Pantry

Store-room

Dining Room

Laboratory

Drawing Room

Library

Storeroom

Floorplan of Priestley's house, first floor, ca. 1800

lived here. He also enjoyed playing backgammon and whist, a popular card game of the time, although he didn't gamble, on principle.

The large kitchen in the Priestley house features a cavernous hearth, where cooking pots were suspended from a crane above the fire and where meat and game were roasted on spits. The constantly burning fire also provided a place where Priestley kept his caged laboratory mice warm in winter, so they wouldn't die from the harsh Pennsylvania cold.

While Priestley was a widely recognized theologian and educator, he is best known for his work as a chemist, in the study of gases. He was one of the great "gentleman" scientists—like Michael Faraday and Benjamin Franklin, he had no formal training in science and was not a professional. And his introduction to the study of gases may seem unusual: He began his experiments at a local brewery in the town of Leeds, England. He discovered that he could produce a pleasant, bubbly drink by combining water with a by-product of beer fermentation now known as carbon dioxide—and that's how soda water, or seltzer, was invented.

Priestley was curious and enthusiastic about his experiments, but he was not very orderly or methodical in what he chose to do. (A great believer in the role of chance in his work, he liked to say that if he had known any chemistry, he would never have made any discoveries.) But he made very careful observations. As a result, as 19th-century chemist Humphrey Davy once remarked about Priestley, "No single person ever discovered so many new and curious substances." The list includes ammonia, sulfur dioxide, carbon monoxide, hydrogen chloride, nitric oxide, hydrogen sulfide, and his greatest discovery, oxygen.

In 1774 Priestley focused light through a burning lens 12 inches in diameter onto a sample of mercuric oxide. In this way, holding the lens 20 inches away from the ore in its flask, he was able to avoid contaminating the mercuric oxide with any other substance, so he knew that whatever was produced came directly from the ore, not from matches, or a gas flame, or any other source. Using these controlled experimental methods, he found

Priestley began working in his Northumberland laboratory as soon as he could. This photo, taken ca. 1982, does not show restorations planned for the late 1990s, but it does show the hood used to vent noxious fumes. (Photograph courtesy of the Pennsylvania Historical and Museum Commission, 1996)

that a new gas, never before discovered, emerged from the ore. The new "air," which seemed somehow purer than normal air, caused a candle to burn brightly and vigorously. He tried the effects of breathing the new air—on mice, on plants, and even on himself. He found breathing it "peculiarly light and easy," and he noticed that mice thrived on it, while plants didn't. In the process, he made early discoveries that eventually led to the understanding of the process of photosynthesis, by which green plants use light to synthesize (produce) carbohydrates from carbon dioxide and water, releasing oxygen as a by-product.

But Priestley didn't call his discovery oxygen. And he did not really realize at first exactly what he had discovered. At the time, a substance that scientists then called "phlogiston" lay at the center of the most popular theory about what happens when wood burns or iron rusts. No one had ever seen this substance called phlogiston, or weighed it. Yet many 18th-century scientists, including Priestley, subscribed to the theory that phlogiston was released or lost by any substance when it burned, calcinated, or otherwise oxidized (although they didn't use that term). For example, tin and lead change when they are heated. The change was caused, according to the theory, by a loss of phlogiston from the original metal.

But there were problems with the theory. If phlogiston was lost in the process, then how did one explain that substances such as tin and lead not only change color when heated but gain weight? The residue that resulted from heating the metal actually weighed more after the heating than the metal did before. Between 1772 and 1774, Antoine Lavoisier in France performed a series of experiments and demonstrations under controlled conditions that finally proved that phlogiston did not exist.

Still, not everyone rushed to agree, and Priestley stuck by the phlogiston theory all his life, calling his new air "dephlogisticated air"—that is, air without phlogiston—a name by which he intended to evoke the way this gas seemed to support life (since it was thought that phlogiston extinguished life).

Oxygen received its name, ultimately, from Antoine Lavoisier, to whom Priestley explained his experiments when he visited France in October 1774. Lavoisier listened with interest and realized suddenly that Priestley had isolated one part of air—that what we now refer to as Earth's atmosphere is made up largely of two gases, one that encourages combustion and respiration and the other that does not. Now the truth seemed clear: Priestley had isolated the gas in the air that supported burning; the new gas that Priestley found was undiluted by the portion of the air that objects did not burn in. In 1779 Lavoisier announced that air is composed of two gases, the first of which—the one that supported burning—he called *oxygen*.

As a result, Joseph Priestley, Unitarian theologian and minister, is most widely remembered for his discovery of oxygen, for which he ranks as one of the great pioneers of chemistry.

In Northumberland, Priestley continued his scientific work, as well as his religious and philosophical writings. He turned down several offers of positions from the Unitarian church in Philadelphia, as well as a professorship in chemistry at the University of Pennsylvania, spending the last 10 years of his life quietly writing on the Susquehanna. Working up to the last, on February 6, 1804, he died quietly in his library at home after dictating some changes in a manuscript to his son Joseph Jr.

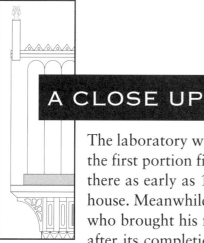

A CLOSE UP   JOSEPH PRIESTLEY'S LABORATORY

The laboratory wing on the north side of Priestley's house was the first portion finished, and Priestley was able to begin work there as early as 1797, while the carpenters finished the main house. Meanwhile, he continued to live with his son Joseph Jr., who brought his family to live with Priestley in his new home after its completion. Priestley made all his own apparatus or had it made under his close direciton, and replicas of some of this equipment reside in the restored laboratory, which was a well-lit room fitted with a fume hood to ventilate noxious gases. In a short time he was able to gather the apparatus for a well-equipped laboratory—an astounding feat, considering the distance and inaccessibility of Northumberland from either Philadelphia or Britain.

Today visitors can view replicas of the apparatus Priestley used in his famous experiment in which he discovered what he called "dephlogisticated

In his later years, Priestley became interested in the science of living organisms and often used his microscope to examine specimens from nearby streams. (Photograph courtesy of Pennsylvania Historical and Museum Commission, 1996)

air"—or oxygen. On display here: the burning lens capable of generating high temperatures by focusing rays of the sun, and the pneumatic trough in which he collected the "air" released from the heated mercuric oxide.

Priestley was also always interested in electricity; in fact he wrote *The History and Present State of Electricity*, published in 1767, which was widely read and served as his introduction to Benjamin Franklin, who was the United States envoy to England at the time. Franklin, who was 27 years older than Priestley, became the younger man's mentor and friend and they maintained a steady correspondence until Franklin's death in 1790. The main instrument used for experimentation with electricity, the electrical machine on display in the Priestley laboratory, was a contraption that created static electricity by means of friction on a rapidly spinning glass sphere. In Northumberland Priestley also began working with the new "voltaic pile," the first electric battery, invented in 1800 by the Italian physicist Alessandro Volta.

The fields and riverbanks near Northumberland supplied Priestley with abundant plant and animal life for his studies, for which he used the microscope that is on display in his laboratory at the Joseph Priestley House. In 1803 he experimented with algae to disprove Erasmus Darwin's theory of spontaneous generation, throwing himself vigorously into this new pursuit. But he was already ill with little time remaining and he never became expert in either botany or zoology. Disproof of the theory of spontaneous generation would have to wait instead until 1859, when it was refuted by French chemist Louis Pasteur.

Priestley also continued writing scientific papers, of which he wrote 30 in the United States, during the last 10 years of his life. His last published scientific paper, which came out in June 1797, was, ironically, "The Doctrine of Phlogiston Established," in which he stubbornly continued to try to defend the defeated phlogiston theory.

The articles and apparatus used by Priestley in his famous researches and discoveries were presented to the Smithsonian Institution in 1883 by his descendants, who were still living in Northumberland at that time.

PRESERVING IT FOR THE FUTURE

After Priestley's death, his house in Northumberland had a succession of owners. In 1874 many of America's leading chemists gathered there to commemorate the centennial of Priestley's discovery of oxygen, and the meeting led to the formation of the American Chemical Society. The Commonwealth of Pennsylvania acquired the house in 1961 and has restored it to reflect the tastes and lifestyle of its original owner. The house and its furnishings reflect the period of Priestley's residence and the site features Priestley's laboratory, as well as exhibits representing his accomplishments and varied interests. The house and grounds are maintained by the Pennsylvania Historical and Museum Commission, which undertook a remodeling of the laboratory wing in 1996–97 to better reflect Priestley's original working environment.

EXPLORING ◆ FURTHER

Books about Joseph Priestley

Davis, Kenneth Sydney. *The Cautionary Scientists: Priestley, Lavoisier, and the Founding of Modern Chemistry*. New York: Putnam, 1966.

Gibbs, F.W. *Joseph Priestley: Revolutions of the Eighteenth Century*. Garden City, N.Y.: Doubleday, 1967.

Graham, Jenny. *Revolutionary in Exile: The Emigration of Joseph Priestley to America, 1794–1804*. Philadelphia: American Philosophical Society, 1995.

Holt, Anne. *A Life of Joseph Priestley*. With an introduction by Francis W. Hirst. Reprint of the 1931 edition. Westport, Conn.: Greenwood Press, 1970.

Related Places

Joseph Henry House
Princeton University Campus
Princeton, NJ 08544

Joseph Henry lived in this two-story brick house while he taught (1832–46) at what later became Princeton University. Henry did significant research in electromagnetism and served as first secretary of the Smithsonian Institution and president of the National Academy of Sciences. Now used by the university, the house has been converted to offices, but the exterior architecture has been preserved.

Benjamin Franklin National Memorial
The Franklin Institute
222 North 20th Street
(at Benjamin Franklin Parkway)
Philadelphia, PA 19103
(215) 448-1200
Internet website: http://www.fi.edu

The Franklin Institute was founded in 1824 to honor inventor Benjamin Franklin, and a larger-than-life statue of the inventor-scientist-printer-diplomat stands in the rotunda. The memorial was dedicated in 1972 and contains exhibits on the life and career of Benjamin Franklin. The Institute also houses an outstanding hands-on museum of science and technology (see More Places to Visit).

McDowell House and Apothecary Shop

HOME OF PIONEER SURGEON EPHRAIM McDOWELL
Danville, Kentucky

AT A GLANCE

Built: about 1795

Home and apothecary shop (1802–30) of Ephraim McDowell, pioneer surgeon

Purchased by McDowell in 1802, this two-story Georgian-style wood-frame and brick house was built in three stages. The original brick ell and adjoining shop building were built in the 1790s. The frame portion was completed in 1804, and a brick addition was finished in 1820. The apothecary shop houses an extensive collection of medical instruments and apothecary ware.

Address:
McDowell House and Apothecary Shop
125 South Second Street
Danville, KY 40422
(606) 236-2804

On December 25, 1809, Dr. Ephraim McDowell performed the first recorded ovarian surgery in the United States. His courage, the success of his technique, and the bravery and subsequent good health of his patient have become legendary.

In his economy he was frugal, his hospitality was informal, with the exception of an occasional children's ball, when fiddler Sam, "in solo," gave them music.
—*Tri Weekly Yeoman* of Franklin County, June 21, 1877

♦ ♦ ♦ ♦ ♦

Ephraim and Sarah Shelby McDowell's House in Danville, Kentucky. Notice the mortar and pestle symbol above the door of the brick apothecary shop on the right. (McDowell House and Apothecary Shop)

In 1783, when Ephraim McDowell's family arrived in Kentucky, it was a sparsely populated area of American wilderness. By the time he had grown, gone away to medical school, and returned 12 years later to establish his medical practice, Kentucky had become a state. But the area was still on the western frontier of the United States, and a walk through Danville's historic sites today brings to mind an era when brave, pioneering people carved out a life at the edge of forested hills, housed themselves with rough-hewn logs, and passionately argued issues of statehood and law over tankards of beer in the town tavern. They wore rough clothes, which they made at home out of materials that they grew and wove, and they either made their own furniture or imported it from cabinetmakers along the Atlantic coastline.

Born in 1771 in the Shenandoah Valley of Virginia, where his father was a land court judge, Ephraim McDowell (1771–1830) was the ninth of eleven children. His Scotch-Irish grandparents had arrived in America earlier in the century. Young Ephraim was 12 when his father was appointed land commissioner and judge in Kentucky, which was at that time a county of Virginia. So the McDowells followed the roads through the wilderness over the Blue Ridge Range and the Appalachian Mountains, through the pass known as the Cumberland Gap, and along the Wilderness Road that Daniel Boone had blazed only a few years before, in 1775. The trip was a frightening one, since the McDowells, along with other Americans of European heritage, were pushing into American Indian hunting territory. Members of the Creek and Cherokee tribes, angered by the intrusion and theft of their lands, had often attacked and killed members of parties like the one the McDowells traveled with.

But Ephraim's family arrived safely and, looking across central Kentucky's rolling green hills today, one can easily imagine the excitement Ephraim must have felt as he began his new life in the country the Indians called *Kentake*, meaning "prairie land." Judge McDowell at first settled his family in Harrodsburg, where a fort had been constructed near the Kentucky River, on the high, fertile bluegrass country plateau. Shortly afterward they moved 10 miles south to Danville, which was fast becoming a political center and, by 1785, was the birthplace of Kentucky statehood.

At 19, Ephraim McDowell left Kentucky to return to Virginia to study medicine. There he apprenticed himself to a prominent physician (American

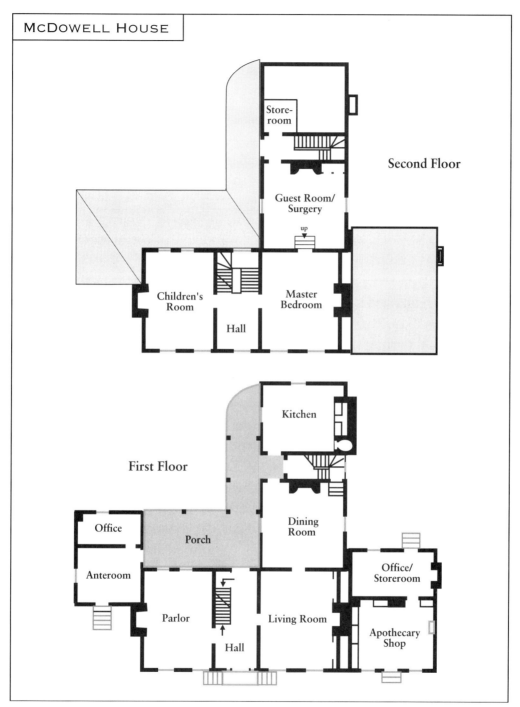

McDowell House

Second Floor

Store-room

Guest Room/ Surgery

up

Children's Room

Master Bedroom

Hall

First Floor

Kitchen

Office

Porch

Dining Room

Anteroom

Office/ Storeroom

Parlor

Living Room

Apothecary Shop

Hall

Floorplan of the Ephraim McDowell House—upstairs and downstairs medical schools were few) and then crossed the Atlantic to study at the University of Edinburgh in Scotland. While there, he studied under John Bell, the most respected surgeon in Europe at the time. Bell discussed diseases of the ovaries (part of the female reproductive system) in detail, and, as it turned out, his lectures proved very useful to McDowell in the future. By 1795, McDowell had returned to Danville and opened his medical practice.

As a young man, Ephraim McDowell was said to have been tall, with "a commanding figure, handsome with black eyes, penetrating gaze, and engaging personality," possessing a refined manner and exceptional intellectual abilities. Moreover, accounts point to his excellence as a conversationalist, being "a ready wit, fond of music, master of the Scottish dialect, sympathetic with tender emotions, and plain and unassuming." In the manner common to the time, he dressed in black, wearing silk stockings to the knee and ruffled shirt fronts.

In 1802, he married Sarah Shelby, the daughter of Isaac Shelby, Kentucky's first governor. And that's when McDowell purchased the brick house and apothecary shop on Second Street, within sight of what was then the town center (and which is now a state historic site).

Today McDowell House is an impressive L-shaped, two-story structure, reflecting the stature achieved by the young surgeon, although he never became wealthy. The house that Sarah and Ephraim moved into in 1802 was the portion that now forms the back of the house—the dining room and the upstairs room that served in 1809 as McDowell's surgery. But by 1803–04 they had added the front rooms. Today, when you step into the parlor at the front of the house, you can almost imagine McDowell writing up his reports or reading at the table. McDowell's medical instruments—or ones similar to those he used—are on display in the dining room, including various surgical instruments (including instruments for drilling holes in the skull to relieve pressure), apothecary jars, and a small medical chest that McDowell purchased in Europe. The master bedroom upstairs is furnished with a washstand, pitcher, and bowl for washing, and a baby cradle. The kitchen, outside and behind the house—separated to prevent the hazard of fire—is especially interesting, with its spinning wheel, open hearth for cooking, lockable sugar cabinet (sugar was an expensive import), butter churn, and sausage mill for making sausage. Most likely, Sarah Shelby McDowell didn't spend much time

The many tasks required for living on the edge of a wilderness centered in the kitchen. (McDowell House and Apothecary Shop)

here herself, though—since she probably had servants to help her manage the household.

By 1809, McDowell was well established as a knowledgeable and capable surgeon. When Jane Todd Crawford in Green County, 60 miles from Danville, consulted her doctor about the painful swelling in her abdomen, he at first told her she was expecting twins. A second doctor concurred. But time passed, the babies had not arrived, and the swelling was getting ever larger and more painful. "She was afflicted with excruciating pain, similar to those of labor, which were incessant," according to one account. The doctors sent for McDowell, who rode from Danville on horseback through winter's cold weather, across the high hills and streams. When he examined Crawford, he concluded "that she was not pregnant; but had a large tumor in the abdomen which moved easily from side to side."

With careful concern, McDowell explained to his patient that he saw no way to improve her condition except through surgery, but he also told her very plainly that the procedure was considered mortally dangerous. In fact, the best information available, as he explained to her, indicated that the danger of inflammation of the abdominal cavity was so great, that the most eminent surgeons in England and Scotland would concur that "opening the abdomen to extract the tumor was inevitable death." However, he was willing to attempt the surgery if she could come to Danville where he could monitor her condition afterward.

Crawford didn't hesitate. A decisive woman, she could see that even the clear possibility of death was worth the risk, since the surgery was her only hope. She rode the 60 miles to Danville on horseback, bruising her abdomen on the saddle horn, she was so swollen.

McDowell chose Christmas Day 1809 to perform the operation. What's remarkable about the surgery is not only that it had never been done before but that, at the time, no anesthesia was available to ease the pain, and no

In this room above the dining room, McDowell murmured a prayer before performing surgery on Jane Todd Crawford, who braved the pain without anesthesia by singing hymns. (McDowell House and Apothecary Shop)

one knew anything about the need for sterile procedures. There were no expert assistants, no intravenous therapy or blood transfusions available. And the risk of infection was great.

But McDowell intuitively kept his instruments clean, carefully drained the wound, and tied off the blood vessels in the enormous tumor to prevent excessive bleeding. He worked quickly, while Crawford sang hymns. The tumor he removed weighed 22.5 pounds. Within five days she was on her feet, making her bed (which apparently concerned her physician), and within 25 days she was on her way back home, "in good health."

It was the first recorded ovarian surgery in the United States, an operation that demonstrated the possibility of safely opening the abdominal cavity. Ephraim McDowell completed numerous other, similar surgeries with success. And his courageous patient, Jane Todd Crawford, lived in good health for another 33 years.

Ephraim McDowell became renowned throughout the region as a fine surgeon, and he performed numerous other successful operations during the course of his career. He continued to practice medicine and live in the house and apothecary shop on Second Street until his death in 1830.

A CLOSE UP   THE APOTHECARY SHOP

Ephraim McDowell maintained his apothecary shop as an integral part of his medical practice, as physicians commonly did in his time. He probably used the back room of the building as his office, or possibly as a storage room, while the front room served as a shop, where patients came to purchase medicines.

There were no prescriptions required, and apprentices would usually fill the orders. Customers might either request a particular herb or remedy, or they might describe the symptoms of an ailment to the physician to ask his advice. A doctor also often traveled to patients' homes, taking medicines along in a medical bag.

The apothecary shop   (McDowell House and Apothecary Shop)

The apothecary shop carried a variety of medicines, such as infusions of digitalis (from a plant, now used to treat heart ailments), plasters, ointments, chalk mixtures for stomach ailments, and balms. The many drawers, bottles, and jars also contained a variety of home remedies, such as mixtures of rhubarb, ginger, senna leaves, catnip, and other herbs. Patented medicines from Germany, England, and France were also sometimes available.

Pills were used sparingly, since they had to be made by hand, a time-consuming task. The apprentice used a mortar and pestle to crush herbs and mix ingredients, as well as a balance to weigh out portions. The McDowell House Apothecary Shop is outfitted with all the necessary equipment, including a mortar and pestle that belonged to Ephraim McDowell, as well as an extensive collection of period apothecary ware. Walking into the shop, you almost expect to see Ephraim McDowell come through the door to discuss your symptoms with you or discuss the politics of the day.

PRESERVING IT FOR THE FUTURE

On May 20, 1939, the restored McDowell house was dedicated as a permanent memorial to this pioneer surgeon, and on August 14, 1959, the restored apothecary shop adjacent to the house was dedicated. Several groups, associations, and corporations have contributed to the restoration and furnishing of the house and shop, notably the Kentucky Medical Association, the McDowell and Shelby families, and the Pharmacists of Kentucky. The site became a National Historic Landmark in 1965 and is maintained by the Kentucky Medical Association. Today it is restored and appointed with furnishings from McDowell's time, and the gardens include trees, shrubs, and herbs of the period.

EXPLORING ◆ FURTHER

Works about Ephraim McDowell

Gray, Laman, Sr., M.D. "After Office Hours," *Obstetrics and Gynecology.* Vol. 16, No. 4, October 1960, pp. 503ff.
———. *The Life and Times of Ephraim McDowell.* Louisville, Kentucky: V.G. Reed and Sons, 1987.

Related Places

Constitution Square State Historic Site
134 South Second Street
Danville, KY 40422-1880
(606) 239-7089

Just a step away from the Ephraim McDowell House (which can be seen through the breezeway of the Constitution Square Jail), this site is the birthplace of Kentucky's statehood. A crossroads for settlers traveling along Daniel Boone's Wilderness Trail into central Kentucky, by 1785 Danville was chosen as Kentucky's first seat of government, and a meetinghouse, courthouse, and jail were built around this square. Today, visitors can still see all of these—constructed of rough-hewn logs—as well as Grayson's Tavern (built 1785), two brick houses (ca. 1816–17 and 1820), and a brick schoolhouse (ca. 1820). A reenactment of the constitutional convention by costumed players takes place each September.

Willis Russell House
204 East Walnut Street
Danville, KY 40422
Mailing address: 216 North Street, Danville, KY 40422
(606) 239-7089 or 236-5315

Originally the home of Revolutionary officer Captain Robert Craddock, this three-story log house was left in 1836 to Willis Russell, a former slave who had become the officer's friend and confidant as well as a teacher. Russell lived here until his death, with his wife and daughter, also taking in six boys, whom he taught, establishing a home school that is believed to be the area's first school for African Americans. Visitors can see original pane windows and flooring dating from ca. 1790. Tours are conducted by appointment.

Pennsylvania Hospital
800 Spruce Street
Philadelphia, PA 19107
(215) 829-3000

Billed as "America's oldest hospital," Pennsylvania Hospital was founded in 1751, with the help of Benjamin Franklin. The first surgical removal of a stone—or mineral mass—from the urinary tract to be performed in the 13 colonies was done here in 1756 by Dr. Thomas Bond. The original building, the Pine Building, with basement cells originally reserved for the mentally insane, is preserved as a museum recalling medicine as it was practiced in the 18th and 19th centuries. Visitors can see plaster anatomical casts used to instruct medical students; the surgical amphitheater, where surgical procedures were performed 1804–68; the Museum of Nursing History; and many historical paintings.

Hopewell Furnace National Historic Site

IRON-MAKING COMMUNITY OF THE 1830s
Elverson, Pennsylvania

AT A GLANCE

Founded: 1771; restored appearance: 1820–40

Founded by iron maker Mark Bird; remained in operation until 1883

An iron-making community near Elverson, Pennsylvania, returned to its ca. 1830s appearance, with 14 restored structures in the core historic area, including a blast furnace, furnished workers' cottages, charcoal hearths, and ironmaster's house.

Address:
Hopewell Furnace National Historic Site
2 Mark Bird Lane
RD 1, Box 345
Elverson, PA 19520
(610) 582-8773 (TDD: 583-2093)

[Here] a succession of energetic ironmasters, furnace workers, and other community members helped lay the foundations of American ironmaking technology.

—National Park Service Handbook 124,
Hopewell Furnace, 1983

♦ ♦ ♦ ♦ ♦

When European colonists arrived on American shores, blast furnaces had been used in Europe for several centuries to make iron, and these settlers brought with them the knowledge and methods for making utensils out of iron ore. To build houses, clear land for farming, and carve out a life for themselves in this new land, the settlers needed axes, saws, picks, hoes, nails, pots and pans, and more, and by the late 1600s ironworks communities sprang up wherever iron was found along the Atlantic seaboard.

By 1771, when entrepreneur Mark Bird founded Hopewell near the headwaters of French Creek in Pennsylvania, iron making not only formed the backbone of American industry but it also had become one of the major issues that fomented the revolutionary break between the American colonies and England in the years leading up to 1776. Colonists wanted the right to the profits to be gained from manufacturing, while England wanted all the colonies' rich ores and raw materials to feed its own hungry factories. And England, eager for the colonies to remain its own ready-made market for finished goods, passed legislation in 1750 to prohibit Americans from making finished iron products. Angered and determined, by the time the War for Independence broke out Mark Bird had Hopewell manufacturing cannon and shot to be used by George Washington's Continental Army.

Hopewell, along with hundreds of other "iron plantations," continued to form the nation's industrial foundation well into the 19th century. The rural landscape became dotted with the tall stone pyramids that breathed flames and smoke, charcoal-fueled iron furnaces that produced the versatile metal so crucial to the nation's growth. Here generations of ironmasters, craftspeople, and workers produced iron goods during war and peace—ranging from cannon and shot to cast-iron stoves and domestic items such as pots and sash weights for windows. By the 1830s Hopewell had become a self-supporting community built on cooperative effort—a group of hardworking individuals who furthered their own interests by contributing to the community enterprise. Shared goals, along with social and family bonds, made plantations like Hopewell into stable and productive communities, the base on which

In this photo you can see the long, sloping roof of the casting house on the left, the company store, the elegant three-story structure of the ironmaster's house in the background on the right, and the blacksmith's shop in the right foreground. (Courtesy National Park Service)

America's iron and steel industry was founded. Hopewell, as well as the iron plantation system, was at its height from 1820 to 1840, the period reflected by the restoration that visitors see when they go there to visit today.

The region around Hopewell had everything needed for iron production: a wealth of iron ore near the surface, limestone for use in removing impurities from the iron, hardwood forests to supply the charcoal used for fuel, rushing water to power the bellows that pumped blasts of air into the furnace fires, and plenty of workers to supply the labor. By the 1830s it had developed a long-held reputation for producing cast-iron stoves, for which there was a steady market. As Pennsylvania added more and more links to its transportation system of roads, canals, and railroads (such as the Main Line Canal and the Allegheny Portage Railroad), parts molded in the Hopewell casting house were shipped to sites all over the East Coast. There they were assembled into stoves and sold from Rhode Island to Maryland as the "Hopewell stove." By the time the last fires burned out at Hopewell in 1883, the little community had produced some 80,000 finely crafted cast-iron stoves.

As you stroll today among the structures of this iron-making community, you may talk with workers of the past, smell bread baking in the open-air brick ovens, view the blast furnace once used to make iron and hear the hissing blasts of air that made its fires burn hot, taste vegetables from the gardens, hear spinning wheels hum, and experience a little of what life was like for those who lived and worked here. You may also see the blacksmith working at his forge—using hammer and anvil to pound hot metal into wrought iron. Or you may observe molding and casting of stove plates like the ones for which Hopewell was famous, as well as the charcoal-making process that the community's blast furnace depended upon. Harvest-time also brings apple picking in the historic orchard that served the community.

Today, however, the quiet of the Pennsylvania countryside demands a little imagination to transform the setting to the noisy, busy scene Hopewell was at its peak. In those days, the furnace stack roared as it spewed out smoke and exhaust; workmen shouted; and the waterwheel clanked monotonously as it turned to pump the furnace bellows. Wagons creaked as they carried ore in from the hills, and a bell in the castinghouse periodically called workers back to work. In the days of its highest production, the Hopewell Furnace blazed night and day. Workers labored in 12-hour shifts to keep it stoked

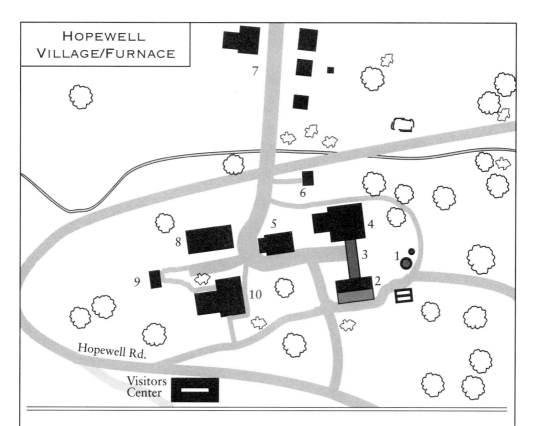

HOPEWELL VILLAGE/FURNACE

1 Colliers used hundreds of charcoal hearths like this one to turn 5,000–6,000 cords (4x4x8-foot stacks) of wood into charcoal fuel each year.

2 The cooling shed was the dumping point for teamsters hauling charcoal, which was often still smoldering. From there it would be moved inside to the adjacent charcoal house.

3 Through the connecting shed fillers carried charcoal, limestone and iron ore to the bridge house at the top of the furnace stack. From there they stoked the furnace. At the base of the furnace below, a waterwheel (not visible) ran the blast machinery.

4 The casting house, which surrounds the furnace stack, provided the area where molders cast iron into stove plates and other products.

5 The company store sold workers goods that were charged against wages.

6 The blacksmith shop provided hardware and horseshoes for the village and was a favorite gathering place.

7 Tenant houses were available for rent from the company by workers. Single workers either boarded with tenant families or across the road at the boardinghouse. Some even boarded at the ironmaster's house.

8 The barn provided shelter for some 36 draft animals and their food.

9 In the springhouse and smokehouse, servants stored and cured foods.

10 The ironmaster's mansion, built in three stages beginning in 1771, served as a focal point for the community.

Charcoal hearth (left) and collier's hut. At similar hearths Hopewell colliers annually converted 5,000 cords of wood into charcoal for fuel. (Courtesy National Park Service)

and filled with ore. Every day the furnace transformed 5 to 7 tons of ore, using 200 shovelfuls of limestone to purify it, and consumed about an acre of hardwood to produce the 720 bushels of charcoal fuel it needed.

Miners in the surrounding mountains, working on contract or on the Hopewell company's payroll, dug up huge quantities of the iron-laden rock that was hauled in to feed the furnace. In the nearby hardwood forests of chestnut, oak, hickory, and elm, as many as 100 Hopewell woodcutters—some of whom were women—chopped down trees, cut them, and split them to be hauled to the coaling areas to be made into charcoal. From these stacks of wood, a group of about 20 colliers (charcoal makers) transformed wood into the hot-burning charcoal used by the furnace. They used a carefully controlled process that involved slowly charring the wood, burning from the center outward in pits covered with leaves and dirt.

Within the community at Hopewell, laborers included fillers who fed the furnace its diet of iron ore, limestone, and charcoal; molders, who cast molten metal into stove plates and other finished pieces; teamsters, who drove the wagonloads of raw materials, supplies, and finished products; and a blacksmith, who hammered out wrought iron and made tools for the mine and furnace.

While the life lived by these laborers seems hard and strenuous—which it was—still, theirs was a better life overall than life was for their city-dwelling counterparts. In Hopewell there were compensations. During the period from 1820 to 1840, the ironmaster at Hopewell was Clement Brooke, a man who had risen up through the ranks of workers at the furnace and knew the work firsthand. For that, he had the respect of his workers, and he, in turn, was concerned about them. Overall, the community had a cohesive sense of purpose and mutual cooperation. Many workers lived in tenant houses furnished by the company, and much of their food was grown on acreage belonging to the company. Whatever they couldn't grow or make for themselves they bought from itinerant peddlers or from the company store,

As in hundreds of other communities that supplied America's iron until 1840, many workers rented company-owned tenant houses like these, gathered around the charcoal burning cold-blast furnaces that provided their livelihood. (Courtesy National Park Service)

where they could charge their purchases against their wages. Unlike city dwellers, they were surrounded by the beauties of the rural Pennsylvania countryside—with its wooded hills, nearby creeks, fields, birds, and wildlife. The community also got together for social activities, such as quilting bees, dances, fairs, and cornhusking parties.

But new technology eventually made Hopewell's furnace outmoded—steel and steam had taken over where iron and water had reigned—and toward the end of the century the heirs of Clement Brooke recognized that their little village and blast furnace could no longer compete in an industrial world based on mass production for a national market. They closed down the furnace in 1883, and with its closing and the closing of other such "iron plantations," a way of life came to an end. These villages were forerunners of the factories and mills of a generation later, and they laid the foundation for the great iron and steel industries on which much of the 20th-century economy became based. But they also were more: They were closely knit villages that sprang up around the resources on which the ironworks depended, a way of life that remains to us as a legacy of our past.

Today visitors can see the company store where workers purchased goods and tour the tenant workers' simple homes, as well as the ironmaster's impressive home, which served as the ironmaster's family home, the business headquarters, a boardinghouse, and a social center.

And as you stroll through the busy grounds, with the blast of air still hissing into the furnace, you can almost feel the heat and see the glow of the molten iron that once formed the heartbeat of this village. And you can readily imagine the thriving community that once called this place both home and workplace.

A CLOSE UP | HOW BLAST FURNACES WORK

The blast furnace at Hopewell no longer burns, but in its day this tall stone structure shaped like a flattened pyramid was the heart of the iron-making community. Hot-burning charcoal was used to melt iron ore, to which workers added a "flux," usually limestone, to combine with impurities, which could then be removed, leaving behind a purified iron product.

The founder was in charge at the furnace, where he supervised a keeper (who supervised the night shift) and two or three fillers—men and boys with

Sand cast molding demonstrations show onlookers how molten iron was shaped into stove plates. In the casting house skilled Hopewell molders produced parts for over 80,000 stoves. (Courtesy National Park Service)

strong arms and backs—who had the hot and dangerous job of feeding the blast furnace with iron ore, charcoal, and limestone. They wore thick leather aprons, boots, and gloves to protect against the heat as they stoked from the tunnel head opening at the top of the furnace.

Inside the stone furnace, flames were intensified by blasts of air from the bellows (hence the name *blast* furnace). Simplified, the carbon in the charcoal combined with the air's oxygen to form carbon monoxide gas. Then the carbon monoxide, which is highly reactive, combined with the iron ore, producing carbon dioxide exhaust that spewed from the top of the stack. The limestone combined with impurities in the iron ore, forming a residue, or slag, which was drawn off from the top of the molten ore.

There were no gauges. By the color and shape of the flame from the chimney and the consistency of the molten metal, the founder judged the moment when the molten iron was ready. Molten iron poured out of the bottom of the furnace into troughs, from which it could either be drawn off to be cast into a product, at the casting house, or poured into crude bars ("pig iron"), which could then be worked in a refinery forge to produce a bar of iron known as "wrought iron." Wrought iron could be forged, then, into many shapes. Or it could be sent through a rolling or slitting mill to produce plates, bars, or nail rods.

The casting arch at the front of the furnace, where the molders worked, was as hot and dangerous as the flaming stack opening. These were the most skilled of the workers at Hopewell, responsible for casting finished products, and they were the most highly paid. At the height of Clement Brooke's period as ironmaster, Hopewell employed 13 to 19 molders for each blast, as well as 7 to 10 molder's helpers.

PRESERVING IT FOR THE FUTURE

The re-creation of Hopewell Village was mandated by the United States Department of the Interior in 1938. The National Park Service set out, at the instruction of Congress, to restore the village and the surrounding area as nearly as possible to the way it looked when the plant was in full operation. Originally, the park encompassed 214 acres, with the addition in 1942 and 1946 of further woodland areas that were central to the mining and ironworks operation, bringing the total to 848 acres. In 1985 the park was renamed Hopewell Furnace National Historic Site. Today the site is largely surrounded by protected lands, both private and public, including French Creek State Park.

EXPLORING ◆ FURTHER

Books about Iron Works

Clark, Mary Stetson. *Pioneer Iron Works*. Philadelphia: Chilton Book Co., 1968.

Provenzano, Richard G., ed. *Remembering Saugus: Growing Up in a Small Town*. Saugus, Mass.: Saugus Historical Society, 1995.

U.S. Department of the Interior. *Hopewell Furnace*. Handbook 124. Washington, D.C.: National Park Service, 1983.

Multimedia Source

Up-to-date visitor information about the Hopewell Furnace National Historic Site and the Saugus Iron Works is available at the National Park Service Internet site: http://www.nps.gov.

Related Place

Saugus Iron Works National Historic Site
244 Central Street
Saugus, MA 01906
(617) 233-0050

Located about 20 miles north of Boston, this quiet park with old-fashioned, weathered-wood buildings recreates a 17th-century iron mill town, reconstructed by the American Iron and Steel Institute as a memorial to the country's early iron industry. As the waterwheel turns, bellows pump, and the blacksmith's hammer rings out, visitors can view the furnace and forge, as well as its rolling and slitting mill, where iron bars were flattened and cut into rods for making nails. Known as Hammersmith, the settlement here was one of the most advanced iron mills of its time (1646–68) and one of the first "integrated" ironworks in North America, where raw materials were converted into finished products in one place. Within 20 years, the iron works at Saugus slowed to a stop as iron ore ran out and competition increased, but it provided the foundation built upon by others, including Mark Bird at the Hopewell Furnace site.

Allegheny Portage Railroad National Historic Site

A "WATERWAY" THAT CLIMBED MOUNTAINS
Cresson, Pennsylvania

AT A GLANCE

Built: 1831–35

The Allegheny Portage Railroad of the Pennsylvania Main Line Canal, in operation 1834–1857

This railroad provided the link between two halves of the Pennsylvania Main Line Canal, originally using horses and later stationary steam engines to pull the canal boats, loaded on flatcars, up the Allegheny Mountains by rope. The Lemon House, a two-story sandstone tavern-house built at the summit 1831–34, provided rest and refreshment for weary travelers.

Address:
Allegheny Portage Railroad National Historic Site
P.O. Box 189
Cresson, PA 16630
(814) 886-6150

Pennsylvania's great canal, the Main Line, was built to bring trade back to Philadelphia that the new Erie Canal had siphoned away. The railroad portage over the Allegheny Mountains connected the two halves of the canal that could not otherwise be linked. The portage was both an innovative technological achievement and a spectacular experience for passengers.

In six hours the cars and passengers were to be raised 1,172 feet of perpendicular height, and to be lowered 1,400 feet of perpendicular descent, by complicated, powerful and frangible [breakable] machinery, and were to pass a mountain, to overcome which, with a similar weight, three years ago, would have required the space of three days. The idea of rising so rapidly in the world, particularly by steam or a rope, is very agitating to the simple minds of those who have always walked in humble paths.

—Philip Nicklin,
1836 passenger on the Allegheny Portage Railroad

♦ ♦ ♦ ♦ ♦

Walking today along the trace of the Allegheny Portage Railway in central Pennsylvania, one sees remnants of two major eras in 19th-century transportation: the great canal boom and the age of the railroad. At the time this unique piece of American history was built, canal construction was at its height and the nation's railroad network was in its early infancy. This singular blend of canal and railroad embodied some of the most intriguing engineering of both eras.

The Allegheny Portage Railroad formed a key part of Pennsylvania's Main Line transportation system, which linked Philadelphia on the east coast and Pittsburgh on the western edge of the state. For two decades the Main Line opened up the regions beyond the Alleghenies and for two decades formed a vital connection between the growing western communities and the East. Along its route flowed agricultural products and raw materials from the Ohio

River region to eastern Pennsylvania. And westward flowed the manufactured goods from the eastern industrial centers.

The idea of the Main Line (formally known as the Mainline of Public Works) began when Pennsylvanians saw that New York State's brand new Erie Canal—completed in 1825—had cut transportation time between Lake Erie and New York City from between 20 and 30 days down to 8 or 10. Freight rates had dropped, too, from $100 a ton to $5. And New York City had suddenly become the most important Eastern Seaboard port—replacing Pennsylvania's Philadelphia. So Pennsylvanians immediately began planning a canal that would compete directly with the Erie.

But the task turned out to be less than simple. The Erie's engineers had just followed the east-west channel of the Mohawk River, but Pennsylvania had no such natural break through the Allegheny Mountains that ran north

Park rangers in 1830s period dress at the Cresson Summit of the historic Allegheny Portage Railroad. The brick structure in the background is the Lemon House. (Courtesy National Park Service)

Canal boat sections were loaded on flatbed rail cars to travel on the Allegheny Portage railroad.
(Courtesy William H. Shank)

to south through the middle of the state. Undaunted, the citizens demanded that their canal stretch all the way across the state. Just how that would be done, they left to the engineers—and they called in a team of the Erie's prime engineers, including James Geddes, Nathan Roberts, and Canvass White. Experienced crews of Irish and German canal builders also arrived on the scene. Ground was broken at Harrisburg alongside the Susquehanna River on July 4, 1826, and the strange route of the Pennsylvania Canal was ceremoniously begun.

By the time it opened in 1834, nine years after the Erie, the Pennsylvania Main Line Canal quickly became news all over the country: The two halves of the canal—the Eastern and the Western Division—followed established river routes on either side of the mountain range. But for the range itself the engineers had decided to use a portage railroad to hoist the canal boats over the mountains—a particularly innovative idea considering that the first railroads began construction in the United States in the late 1820s and no locomotive yet built had enough power to scale the mountain slopes.

The engineering used at the Allegheny Portage Railroad was straightforward: hoist an object over an obstacle using pulleys and ropes. But the execution was uncommonly bold: The obstacle was a major mountain range, and the objects to be hoisted were canal boats, divided in sections and loaded on flatcars. The Allegheny Portage Railroad included ten inclined planes, like giant stair steps, five going up one side of the mountain and five more going down the other side. Passengers and goods were moved to railroad cars to travel 36.6 miles of their journey. For 6.6 of those miles stationary engines and ropes pulled them up one side of the mountain and lowered them down the other.

During the portage's busiest periods, six trains an hour were pulled up each incline on the Allegheny Portage Railroad. When possible, the operators used cars descending on the other track to counterbalance those ascending, lessening the strain on the engines.

By today's standards, these were not particularly steep inclines. The steepest—number 8—had a slope of 9.9 percent (a 9.9-foot vertical rise in 100 horizontal feet), or less than 6 degrees. The average incline rose about an inch every foot—which would barely be considered a hill in San Francisco or on the slopes of the Rocky Mountains. But they were too much for low-powered early steam locomotives, although they were designed to be easy enough for horses to pull one car at a time up the slope if the stationary engines broke down.

Once the cars reached the canal basin at Johnstown, at the foot of the mountains, the boat sections were once more eased into the water, joined back together, and floated down the canal's Western Division to Pittsburgh and the Ohio River.

Travelers from Philadelphia who wanted to visit relatives in Pittsburgh by way of the new canal system set out early, not on a boat but on a "State Railroad Coach." Drawn by horses along the tracks, the "railroad coach" would proceed at what was then a relatively fast pace. Occasionally a hill would be so steep that the horses were unhitched and the coach was hooked onto a cable that would pull it up the tracks to the crest of the hill. Once at

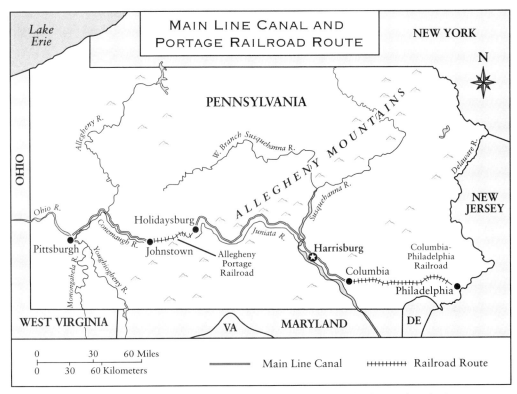

Map of the Pennsylvania Main Line, showing the Columbia-Philadelphia Railroad, the Eastern Division of the Main Line Canal from Columbia to Hollidaysburg, the Allegheny Portage Railroad over the mountains, and the Western Division from Johnstown to Pittsburgh.

the top travelers would wait until the cable was unhooked and a fresh team of horses was hitched up. Then they would be on their way again, moving by rail down the narrow track another 75 miles or so until they reached Columbia.

At Columbia, weary after the 20-hour journey from Philadelphia, they would be shifted to a canal boat. This boat traveled in a "great ditch" dug alongside the Susquehanna River, traveling uphill through locks—mechanisms that raised and lowered boats to canal sections of different heights. On this trip the boat traversed 43 miles, until it reached a tiny artificial lake created by a high dam where the Juniata River flowed into the Susquehanna. There travelers would find themselves and the boat being towed, still traveling on canal water, through a series of aqueducts, across the artificial lake and the Juniata River. After passing through another 88 locks to achieve a rise of 584 feet, the travelers on the canal boat would then continue another

127 miles on to Hollidaysburg, Pennsylvania, where at last they could disembark for the night before the next leg of the journey.

In the canal basin at Hollidaysburg, the portable boat sections in which some passengers had traveled from Philadelphia were floated onto railroad cars for the portage. They were hauled from the water by stationary steam engines, then pulled by locomotives at about 15 mph over the long grade to the first incline. In a small shed at the foot of the incline, workers hitched three cars at a time, each with a load averaging 7,000 pounds. This cable was pulled at about 4 mph by a stationary steam engine beneath a large shed at the top of the incline. (The railroad also had ordinary railroad coaches—also pulled up the incline—for those traveling aboard ordinary canal boats.)

If the gently rocking canal boat had lulled travelers to sleep, the Portage Railway was quick to wake them up again. Once more the horses were unhitched and the boat or coach was attached to another cable, as travelers were suddenly confronted with the fact that they were now going to be pulled up the tracks for a total climb of some 1,400 feet (heading in the westward direction). The motor-driven cables would pull the passengers up the first of five slopes. At the top of each incline, the boat or coach would switch once again to natural horsepower, to be pulled to the beginning of the next incline. There it would be attached to yet another cable, pulled steeply upward along the tracks, unhooked, and hitched up to horses again. The process would be repeated three more times until the summit of the mountain was reached. There exhausted travelers would disembark for a cup of tea or a swallow of ale, perhaps even a quick game of backgammon, at the Lemon House at Cresson Summit.

Then they would board and repeat the entire process as they were let down the opposite side of the mountain, where they traveled through the longest tunnel then in America, 901 feet of semi-darkness, finally emerging from the six-hour, 37-mile portage at Johnstown. Here they would board another canal boat to begin the concluding leg of their journey through a 105-mile section of canals, including another series of locks—and on at last to Pittsburgh.

Although slow and fatiguing, the trip certainly was a colorful experience —doubtless a good conversation starter upon arrival and for years to come. Even the English novelist Charles Dickens traveled the Portage Railroad, which he described in eloquent terms, writing in 1843, "Occasionally the rails are laid upon the extreme verge of the giddy precipice and looking down from the carriage window, the traveller gazes sheer down without a stone or scrap of fence between into the mountain depths below."

Unfortunately, even though the fast packet boats (canal boats) made the trip in six days—competitive with the Erie Canal—things had already begun to change by the time the Main Line was finished. Businessmen in a hurry and farmers and traders shipping freight were not very happy with the time or the expense involved in traveling Pennsylvania's new canal. Branches were gradually added, forking from the Main Line Canal, but the Pennsylvania system was not a financial success. Its operational costs were much too high and could never pay off its initial debts. Moreover, the Pennsylvania Railroad had begun laying tracks across the state, and after 1852, traffic on the Allegheny Portage Railroad quickly faded. In 1857 the Pennsylvania Railroad purchased the Main Line from the Commonwealth of Pennsylvania and traffic on the canal and the portage railroad came to an end.

Engine House #6 at Cresson Summit   (Courtesy National Park Service)

Today visitors to the Allegheny Portage Railroad National Historic Site can view this piece of transportation history from atop the Allegheny Mountains. At the visitor center, exhibits show what the railroad was like in its heyday, and a film tells the story of the portage railroad. From there a boardwalk leads through a stone quarry to Inclined Plane Number 6, and visitors can follow the Incline 6 trail to the Skew Arch Bridge, a bridge built at an angle. A second interpretative trail takes visitors along the summit level. By reserving a place on formal ranger-guided hikes during the summer months, visitors can also view incline planes 8, 9, and 10, stone culverts, and the Staple Bend Tunnel, the first railroad tunnel constructed in the United States.

On self-guided tours, visitors can also explore the Engine House 6 Exhibit Building, which preserves the remains of the original engine house foundation and features a full-scale model of the stationary steam engine that pulled the portage railway cable in later years.

Replicas of the rails used on the Allegheny Portage Railroad. Wooden stringers were fastened to cross-ties, then fitted with iron strap rail on the inside edges for durability. (Courtesy National Park Service)

Stone sleepers, also known as sills, were cut from local sandstone to form the railroad structure. Workers cut more than 193,000 of these blocks to support the rails. (Courtesy National Park Service)

Walking on the grounds of the Allegheny Portage Railroad National Historic Site, you can imagine the bustle of weary and excited travelers as they experience the new freedom of traveling, not on foot or horseback, but aboard cars transported by rail and water. You can imagine the squeal of the steel wheels on the rails, the rumble of the cars as they ascend and descend the slopes, the racket of the steam engines hauling the cables. You half expect to see women and men straightening their clothes as they descend from the carriages and hurry into the Lemon House to refresh themselves with a bit of tea or sip some sherry or ale. You can almost hear their animated conversations around you.

And you can appreciate the effort, thought, and skill that went into building the Allegheny Portage. Costumed log hewers demonstrate how logs were cut and hewn to construct the rails, which were made of wood and then covered with iron. At the stone quarry you can see how stone was quarried and cut to build the railroad, much of which was built on sandstone "sleepers," which anchored the cross ties. And rangers tell what it was like to work on the railroad, as well as to travel on it.

In addition to a ranger-narrated bus trip along the entire 37-mile path of the Allegheny Portage, guided hikes are scheduled throughout the summer months, alternating among a half dozen excursions. (Call or write for a schedule.) Hiking destinations include the Staple Bend Tunnel (4-mile hike); Incline 8, the longest and steepest of the 10 inclined planes (3,117 feet, with a rise of 308 feet); Canal Basin at Hollidaysburg, including a walking tour through the historic town; a tour of Inclines 6 to 10, a 7-mile hike of the eastern slope of the portage; and a hike to Incline 10, the easternmost incline.

Nearby, an exploration of the Lemon House Inn recreates the aura of a travelers' haven of bygone days. Here weary travelers stopped to rest, drink, and eat. The inn was built by Samuel Lemon about 1832 and served as his home, tavern, and business office.

A way station in time, the Allegheny Portage Railroad not only evokes an era when travel was far slower and less comfortable than today but also celebrates the ingenuity of people faced with what seemed an insurmountable obstacle.

A CLOSE UP   THE LEMON HOUSE

According to tax records, by 1826, at the age of 31, Samuel Lemon had bought 288 acres of land near the summit at Blair's Gap in Cambria County, Pennsylvania. He cleared the land and opened several stone quarries on its upper levels, and in 1831 he was delighted to hear that the Main Line Canal Commission had decided to build a portage railway for the Main Line Canal right past his property. Immediately he set about constructing a sturdy inn to replace the log cabin he had. He used hand-hewn sandstone, cut from the nearby quarry you can still see today, to build the two-story structure. And

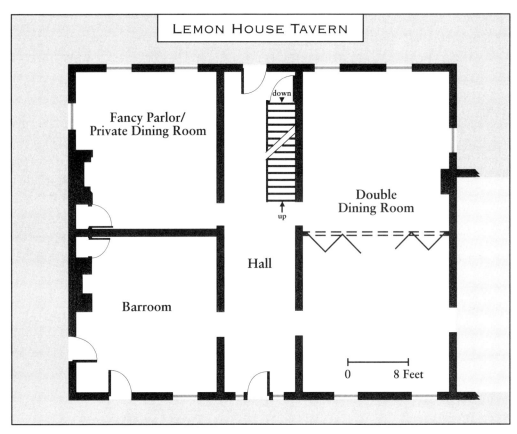

Fancy Parlor/
Private Dining Room

down ▼

Double
Dining Room

up

Hall

Barroom

0   8 Feet

Floor plan of the first floor of the Lemon House tavern at Cresson Summit (Courtesy National Park Service; redrawn by Jeremy Eagle)

he set aside the three large rooms off the main hall on the first floor for his wayside tavern business.

Walking into this building today gives us a glimpse into the dreams and visions, as well as the customs, of another time. Here, Samuel Lemon and his wife, Jean Moore Lemon, built an enterprise and made a good living based on a mixture of good luck and hard work, and their inn stands as a symbol of the Allegheny Portage Railroad itself—a tribute to the idea that people often find a way around obstacles when getting somewhere is important enough to them.

Today, as the Lemon House stands at the summit of the range crossed by the portage railroad, we can imagine the euphoria of travelers arriving from their long journey at this pinnacle stopping point in their voyage. As they descended from the train, they would stretch their legs, feel the summit winds ruffle their clothes, and walk over to the Lemon House (or the Lemon Mansion House as it was often called) for a warm meal or refreshing drink.

To the left off the main hall, in the barroom, male travelers would hang their coats on pegs, drink a tankard of ale at the bar, or talk or play cards around the tables. This was the main center of activity of the building—for men, at least—except at meal times. Conversation might turn to politics, business, or boisterous joking. Women generally did not go into this room, and certainly not women of higher social standing. This was a place where the men didn't have to worry about "parlor manners."

Across the hall, visitors can walk into the double dining room. Double doors could be closed to separate the two halves of the room, which provided a common dining area for everyone—passengers and crew of the portage

Ladies usually retired to the fancy parlor at the Lemon House to rest from their journey. (Courtesy National Park Service)

railroad, as well as other travelers and wagon drivers. The furniture is solid and serviceable, with tables large enough to seat six guests at a time and sturdy enough to stand up to heavy wear. Meals served here were quickly consumed, during train stops, since most travelers didn't spend the night.

Ladies traveling in the company of gentlemen would generally go to a third room, however—the fancy parlor or private dining room down the hall to the left. Here they might play a quiet game of backgammon, sip tea, and warm themselves by the fire. The furniture in this carpeted room was nicer, the atmosphere more genteel, and the effect was relaxing after a noisy, exciting ride to the summit along one of America's pioneer railways.

PRESERVING IT FOR THE FUTURE

Designated a unit of the National Park Service in 1964, the Allegheny Portage Railroad National Historic Site is administered by the National Park Service, U.S. Department of the Interior. The National Park Service preserves remnants of the railroad, including portions of four inclined planes, the Lemon House, and the Staple Bend Tunnel.

EXPLORING ◆ FURTHER

Books about Canals and the Allegheny Portage Railroad

Gies, Joseph, and Frances. *The Ingenious Yankees*. New York: Thomas Y. Crowell Co., 1976.

Shank, William H. *The Amazing Pennsylvania Canals*. York, Penn.: American Canal and Transportation Center, 1981.

———, ed. *Sylvester Welch's Report on the Allegheny Portage Railroad, 1833*. With a 1988 introduction by W. H. Shank. Gettysburg, Penn.: Thomas Publications, 1975.

———. *Three Hundred Years with the Pennsylvania Traveler*. York, Penn.: American Canal and Transportation Center, 1976.

————, et al. *Towpaths to Tugboats: A History of American Canal Engineering,* Second Edition. York, Penn.: American Canal and Transportation Center, 1991.

Related Places

Johnstown Flood National Memorial
P.O. Box 355
St. Michael, PA 15951
(814) 495-4643

Located a short distance from the Allegheny Portage Railroad site, this memorial mourns those lost on May 31, 1889, when an earthen dam collapsed, causing a flash flood of the city of Johnstown about 14 miles down the Little Conemaugh River. More than 2,000 people died in the flood, which destroyed the once prosperous industrial center of Johnstown.

California State Railroad Museum
111 "I" Street
Sacramento, CA 95814
(916) 445-7387

This 15-acre museum houses the largest interpretive railroad museum in the nation, with carefully restored locomotives and cars from the 1860s to the 1960s. Visitors can climb aboard steam locomotive cabs and walk through a Pullman sleeping car from the 1940s. They can also explore passenger coaches from the turn of the century, an icebox car, and a freight car. The museum runs an authentic steam excursion railroad, the Sacramento Southern, which visitors can ride for six miles along the Sacramento River on weekends from April through September. Diesel-powered trains run on a limited weekend schedule October through December. Locomotives from the museum's collection pull the train of open and closed excursion cars.

Chesapeake and Ohio Canal National Historical Park
Headquarters
Box 4
Sharpsburg, MD 21782
(301) 739-4200

This park preserves the 184.5-mile C&O Canal, which operated from 1828 to 1924 along the route of the Potomac River between Washington, D.C. and Cumberland, Maryland. Hundreds of original structures—including locks, lockhouses, and aqueducts—serve as reminders of the canal's key transportation role during the Canal Era. The park has five visitor centers:

Georgetown:
1057 Thomas Jefferson St., NW
Washington, DC 20007
(202) 653-5190

Great Falls Tavern:
11710 MacArthur Boulevard
Potomac, MD 20854
(301) 299-3613

Williamsport:
205 W. Potomac Street

Williamsport, MD 21795
(301) 582-0813

Hancock:
326 E. Main Street
Hancock, MD 21750
(301) 678-5463

Western Maryland Station:
Canal Street
Cumberland, MD 21502
(301) 722-8226

George Washington Carver
National Monument

BIRTHPLACE OF BOTANIST/AGRONOMIST
GEORGE WASHINGTON CARVER
Diamond, Missouri

AT A GLANCE

Built: Birthplace site, originally ca. 1855;
Moses and Susan Carver house, 1881

Home of botanist/agronomist George Washington Carver, ca. 1865–1877

Located near Joplin, 210 acres of the original 240-acre plantation farm
on which former slave George Washington Carver spent his boyhood.
Surrounding woodlands and prairie grasslands are preserved, as well as his
partially reconstructed birthsite and the small clapboard home belonging to
Moses and Susan Carver, the couple who took him in when he was orphaned.

Address:
George Washington Carver National Monument
5646 Carver Road
Diamond, MO 64840
(417) 325-4151

My work, my life, must be in the spirit of a little child
seeking only to know the truth and follow it.

—George Washington Carver

♦ ♦ ♦ ♦ ♦

When we want a good, solid, reliably nutritious quick-to-fix lunch, the American standby is the peanut butter sandwich. In fact, Americans eat more peanut butter than the citizens of any other nation in the world—most other nations use peanuts primarily for oil. And the story of how this happened began on this farm in southeastern Missouri, just as the Civil War drew to a close.

Before the Civil War, only the poorest people in the South ate peanuts. They fried them or ate them boiled, much as we do peas (of which they are a cousin—both are part of the legume family of plants). During the war, soldiers on both sides ate what they called "goober peas," another name for peanuts, but after the war, the taste for peanuts tapered off without completely vanishing. Then P. T. Barnum began selling roasted peanuts at his circus in the 1880s, putting peanuts on the snack food list for the first time. And in 1880, a physician in St. Louis wanted a high-protein food to recommend to his patients, so he invented peanut butter. But no one was really *growing* peanuts as a commercial crop until a scientist-inventor named George Washington Carver (ca. 1864–1943) came along and saw that the peanut had a great deal to recommend it, as well as many uses (300 of which he found during his lifetime).

When George Washington Carver was born on Moses and Susan Carver's plantation around 1864 or 1865 (there are no records), the South was still at war against the North, and the future scientist was born into slavery. He began life in a one-room log shanty, which has been partially reconstructed. Moses Carver had purchased George's mother, Mary, when she was 13. George's father, who lived on a farm nearby, was killed in an accident about the time the boy was born. "I am told that my father was killed while hauling wood with an ox team," Carver once remarked. "In some way he fell from the load, under the wagon, both wheels passing over him."

Today, the George Washington Carver National Monument's park sweeps across 210 acres of rural southwestern Missouri, about 15 miles southeast of Joplin. It is a land of prairie grasses dotted with shaded woods and shallow streams. In this gentle oasis of singing birds, wildflowers, and butterflies, the future botanist and agronomist developed a keen fascination with plants. As you walk the ¾-mile nature trail that winds its way through the woodlands and tall-grass prairie, you can see not only the place where Carver was born

The George Washington Carver National Monument preserves the natural beauty that Carver grew up with. (Courtesy National Park Service)

Carver as a boy, sculpted by Robert Amendola (Courtesy National Park Service)

but also the statue by Robert Amendola of Carver as a boy. You also see Carver Spring, Williams Pond and its spring, the house Moses and Susan Carver moved into in 1881 (although George never lived there), and the Carver family cemetery.

In the words of park ranger Lana Henry, "Just being on the farm and spending 10 to 12 years wandering through the woods, we believe this environment influenced him to become what he did later in life."

Carver was born just before the Civil War ended, at a time when extreme turmoil reigned in Missouri. Although Missouri had sided with the Union, it had been admitted as a slave state in 1821, under the Missouri Compromise. Pro-slavery feeling remained high, and although Moses and Susan Carver owned slaves, they were Union sympathizers who opposed slavery in principle if not in practice. And because of their position, which was known in their region, they came in for heavy harassment from raiders and bushwhackers, or guerrilla fighters, that roamed the area. (Many of the secessionists who had been outvoted when the war broke out had retreated to Neosho, Missouri, only 8 miles away.) The Carvers had already been attacked twice when a band of raiders swung through their plantation to make yet another attack. Moses saw them coming and was able to rush George's five-year-old brother Jim to safety. But he couldn't reach Mary and her baby, who were captured and abducted.

Moses talked to a neighbor who thought he might be able to track and find them, and the man did find George, cast aside in a field. But Mary was never found. The Carvers, who did not have children, took the boys in, raised them in their house, and gave them their last name. Jim and George grew up together, as Carver later related, "sharing each other's sorrows," never knowing what happened to their mother.

As a child, George was too weakened and frequently ill to work in the fields, so he did chores around the house and garden. He seemed to have a knack with growing things, even as a boy, and the plants in the garden thrived under his care. Soon he became known around the area as the "Plant Doctor."

But more than anything else he loved the time he spent in the woods. He began a garden of his own in the woods, transplanting and cultivating plants. He observed what made them healthy and what made them grow. "I wanted to know every strange stone, flower, insect, bird, or beast," he later recalled. He began a collection of objects he found—stones, reptiles, plants, frogs, insects. Finally, Susan Carver, exasperated with these "treasures," began to ask him to empty his pockets at the door.

Carver's curiosity went beyond the natural world, though. He wanted other knowledge as well. Susan Carver began teaching him to read, and by 1876, the Carvers found a private tutor for him, but he soon longed for more knowledge.

So by 1877, he was able to convince the Carvers that he was old enough to attend the school in Neosho, the county seat, where there was a school for African Americans. In Neosho he found a black couple, Andrew and Mariah Watkins, who let him stay with them during the week in exchange for doing chores, and he visited the Carvers on weekends. But the teacher had less to offer than Carver had hoped for, and in the late 1870s, George Carver wandered northwest to Kansas, hitching a ride to Fort Scott in 1878. There, he witnessed a lynching for the first time and—aghast at the wanton killing of a man just on suspicion, without a trial and without a just process of law, simply because he was black—George Washington Carver left town that night.

Carver settled elsewhere in Kansas long enough to graduate from high school, which he did in 1885, and in 1890 he entered Simpson College in Indianola, Iowa. Talented as an artist, he considered becoming a painter but decided instead to pursue a career in agriculture, transferring in 1891 to Iowa State Agricultural College (now Iowa State University) in Ames, Iowa. He was the first African American to attend Iowa State, as well as the first black student to graduate, which he did in 1894. But Carver wasn't satisfied with a bachelor's degree. He went on to earn his master's degree in 1896, specializing in bacteriological laboratory work in systematic botany. He soon took charge of both the greenhouse and bacterial lab, also becoming the first black member of the Iowa State faculty.

Never ordinary in any sense, Carver had extraordinary self-reliance as a young man. And although his manner was humble, modest, and frugal—traits for which he is sometimes criticized—he was driven by an inner force to expand his knowledge and increase his understanding. He set out to secure a graduate-level degree at a time when very few people even considered pursuing an undergraduate education. His work always came first. He lived simply and alone, a Spartan existence, with no concern for clothing or other needs that most people consider basic. Yet he was open, warm, and charming. He disliked treading on others' toes, and tended to accommodate rather than confront, but he often used his sense of humor to make his convictions known. A deeply complex man, he exhibited both a strong ego and a self-effacing manner. His outlook was both scientific and religious, and he was at the same time demanding and giving.

From early boyhood, Carver had two traits that stayed with him all his life and shaped his work: love of nature and frugality. He was fascinated with nature and could "see" where others only "looked." This attentiveness enabled him to perceive hitherto unseen possibilities even in the most familiar components of daily life. He developed a methodology that relied on observation first, then experimentation. And an ingrained abhorrence of waste led him to find uses for what seemed useless.

In 1896, Booker T. Washington, the head of Tuskegee Institute in Alabama —a well-known black college—invited Carver to join the faculty there. Carver's talent as a teacher quickly came to light, and he soon became known as the "Wizard of Tuskegee," not only because of his efforts to help his students, but also because of his successes in extending the school's influence to farmers in the surrounding area.

Everyone in the South who was dependent on farming—including a very large proportion of the black population—was experiencing the effects of decades of taking nutrients out of the soil without returning them. Year after year for the past 100 years, farmers in this area had been planting crops that robbed the soil of its nutrients—notably cotton and tobacco. Now they were suffering from poor crops and blighted plants. Carver began talking about alternative crops without making much headway. And then in 1914, the boll weevil came along and destroyed field after field of cotton. Finally, especially

Carver had a great knack for growing things. (Courtesy National Park Service)

in Georgia and South Carolina, farmers began to recognize that an alternative crop was the only solution that could save their farms. And, at Carver's instigation, they began to turn to peanuts.

Carver helped found the department of agriculture at Tuskegee, and for the next 47 years he focused his research and his efforts on persuading cotton farmers to improve their farms by rotating crops—planting crops that enriched the soil instead of depleting it—and using natural fertilizers to restore worn-out soil.

Carver used creative ways to get ideas for improved farming practices across to southern farmers—he was particularly interested in black farmers. He set up conferences and a traveling school with exhibits (which he called his "school on wheels") to teach Alabama farmers the basics of soil enrichment. He held demonstrations and gave public lectures. He also directed a State Agricultural Station.

In 1910, Carver became head of the newly created Department of Research at Tuskegee, and he introduced a new science, known as chemurgy, which is the development of industrial uses for agricultural products. For example, you can make wall board from pine cones, synthetic marble from sawdust, and woven rugs from okra stalks. Carver produced more than 500 dyes from 28 kinds of plants.

But it was Carver's work with the peanut that made him an American legend. One day a farmer came to him asking what to do with the peanuts rotting in the fields and storehouses. So Carver set to work taking the peanut apart and putting it back together again.

What interested Carver most about the peanut was its potential for reversing soil depletion brought on by the over-planting of cotton in the South. If he could find uses for it, a market could open up, and farmers could have a strong cash crop that also would improve the soil on their farms. So he set about finding products that could be made from "the humble goober." His results were dazzling. He developed over 300 products from the peanut, among them: milk, cheese, butter, cream, coffee, vinegar, salad dressings, soap, face powder, plastics, paper, printer's ink, shampoo, insulation, wood stains, linoleum, and cosmetics. He also stressed the nutritional value of peanuts, especially their high protein content.

He literally rescued the South's economy and is often credited with freeing the South from its unhealthy dependence on cotton.

In 1910, *Carver became head of the Department of Research at Tuskegee Institute in Alabama.*
(Courtesy National Park Service)

By 1918, the United States Department of Agriculture was consulting Carver regarding ways to overcome wartime food shortages. At the request of the United Peanut Association, he testified before the congressional Ways and Means Committee in 1921.

When George Washington Carver came before the powerful Ways and Means Committee of the United States House of Representatives, the representative chairing the meeting told him he had 10 minutes. Testifying at the request of the American peanut industry, he was speaking on behalf of placing a tariff on imported peanuts as a protection against foreign competition.

As usual, Carver had paid no attention to dressing for the occasion. He had on his customary old, wrinkled suit, although he wore a flower in his lapel. In his high-pitched voice he explained that he was an agricultural scientist engaged in research in Tuskegee, Alabama. But as he put the products he had developed on the table, he began to captivate his audience. He was knowledgeable, witty, and persuasive. Before long, the committee waived the time limit and the Chair declared, all trace of antagonism gone, "Go ahead, brother. Your time is unlimited."

Carver had brought samples of candies made from peanuts, and breakfast cereals, Worcestershire sauce, and peanut milk, even instant coffee. The testimony lasted almost an hour. His expertise and his persuasiveness had won over one of the most formidable and powerful political bodies in the country. It was a moment that brought him great fame—and his name became a household word.

Unfortunately, during Carver's lifetime the press played up his success to illustrate the accomplishments possible for black members of a segregated society. In more recent years, the fact that his story was used in this negative way has sometimes obscured the striking story of his pioneering work in his field and his success as a scientist, educator, and agricultural reformer. His dedication to his work continued to the end of his life. In 1935 he became a collaborator in the Division of Plant Mycology and Disease Survey of the Bureau of Plant Industry of the U.S. Department of Agriculture. And in 1940 he donated all his savings to the establishment of the George Washington Carver Foundation at Tuskegee for research in natural science.

During his lifetime George Washington Carver received many honors, doctorates, and medals. In 1923, the National Association for the Advancement of Colored People (NAACP) awarded him the Spingarn Medal for his accomplishments, and he received an honorary doctorate from Simpson College in 1928. The Royal Society of Arts in London, England elected him a member.

On January 5, 1943, Carver died in his sleep. Although he had no known relatives (Jim had died as a young man), people from all over the world came for three days to pay him tribute. In April 1990, in honor of his lasting importance to our heritage, he was inducted into the National Inventors' Hall of Fame in Akron, Ohio.

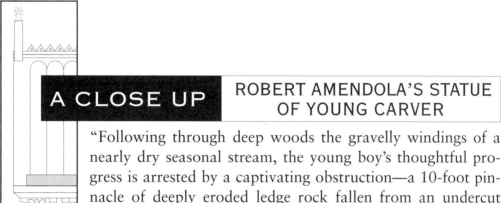

A CLOSE UP ROBERT AMENDOLA'S STATUE OF YOUNG CARVER

"Following through deep woods the gravelly windings of a nearly dry seasonal stream, the young boy's thoughtful progress is arrested by a captivating obstruction—a 10-foot pinnacle of deeply eroded ledge rock fallen from an undercut outcropping in the side of the ancient river bank to land point up in the middle of the stream which had shaped it." So sculptor Robert Amendola once recounted the scene that he imagined when he sculpted the Boy Carver statue that now resides about 130 yards up the trail from the visitor center at the George Washington Carver National Monument.

The figure in Diamond, Missouri is actually the second of Amendola's Boy Carver statues. The original sculpture, designed in the 1950s for the New York Housing Authority, took Amendola two years to complete. It was cast in bronze and erected in a children's playground area for the Carver Housing Development on 101st Street, between Park and Madison Avenues in New York City. Not until seven years later, on a visit to Diamond, did Amendola realize that the statue would find perfect harmony there, as well.

"From early youth," sculptor Rober Amendola wrote of Carver, "he was impelled by a burning curiosity to daily set aside time for a walk alone in the wood, seeking out and contemplating some facet of nature while communing with its creator—a practice he continued throughout his long life. It is one such moment in his boyhood this sculpture means to commemorate."

For more than 25 years, Amendola also used a small scale model of the statue to work with blind people. He used the model to demonstrate that, even without sight, we have capabilities for form perception through touch and our sense of movement. Some 2,000 people used the model in the course of their rehabilitation training program. The much-handled plaster model is now in the St. Louis Museum of Science and Natural History in St. Louis, Missouri.

In 1991, a third sculpture of the Boy Carver was installed in Indianola, Iowa, at Simpson College, which George Washington Carver attended as a young man.

But the figure seems especially at home in the natural setting where Carver grew up. For many visitors to Diamond, Missouri, the statue of a boy deep in thought has become a focal point of their visit. The lean bronze figure seated on a large boulder seems to represent the early development of Carver's special relationship with nature and his unique understanding of his surroundings, capturing the youth, vitality, and sensitivity of Carver as a young boy.

PRESERVING IT FOR THE FUTURE
◆

Created under an Act of Congress in 1943, shortly after Carver's death, the George Washington Carver National Monument was dedicated on July 14, 1953. The park consists of 210 of the original 240-acre Moses and Susan Carver farm and it includes a park visitor center featuring films about Carver's life. A museum diplays exhibits of his scientific research, his artwork, and other artifacts, in addition to the partially reconstructed log cabin at the site where he was born, Moses and Susan Carver's plantation house, the Carver family cemetery, and a bronze statue of the scientist as a boy. The monument is administered by the National Park Service.

The house built by Susan and Moses Carver in Diamond, Missouri in 1881 was small and modest.   (Courtesy National Park Service)

EXPLORING ◆ FURTHER

Books about George Washington Carver

Adair, Gene. *George Washington Carver, Botanist.* New York: Chelsea House, 1989.

Carver, George Washington. *George Washington Carver in His Own Words.* Edited by Gary R. Kremer. Paperback edition. Columbia: University of Missouri Press, 1991.

Coil, Suzanne M. *George Washington Carver.* New York: Franklin Watts, 1990.

McMurry, Linda O. *George Washington Carver: Scientist and Symbol.* New York: Oxford University Press, 1981.

Video Recording

Fabian, Rhonda, and Jerry Baber, producers/directors. *George Washington Carver* [30-minute videorecording, VHS format]. Bala Cynwyd, Penn.: Schlessinger Video Productions, 1992.

Related Place

Tuskegee Institute National Historic Site
George Washington Carver Museum
1212 Old Montgomery Road
Tuskegee Institute, AL 36088
(334) 727-3200
Mailing Address:
P.O. Drawer 10
Tuskegee Institute, AL 36087

The historic site is as much a tribute to Carver as it is to Booker T. Washington, who founded the Tuskegee Institute, where Carver spent most of his career serving on the faculty. The George Washington Carver Museum and Research Center, located on the campus, in particular commemorate his work.

Lick Observatory

MOUNTAINTOP ASTRONOMY'S FIRST HOME
San Jose, California

AT A GLANCE

Built: 1887

The first mountaintop observatory

High on the peak of Mount Hamilton, this astronomical observatory was built to house what was then the world's largest telescope, the Hamilton 36-inch refractor telescope.

Address:
Lick Observatory
UCO/Lick Observatory
Mount Hamilton, CA 95140
(408) 274-5061
http:// www.ucolick.org

San Francisco real estate entrepreneur James Lick established a trust in the 1870s to construct and erect "a powerful telescope, superior to and more powerful than any telescope yet made, . . . and also a suitable observatory connected therewith."

It has been said that the degree of civilization attained by any nation may be estimated from the provision made by its government and people for the study of the stars.

—W. W. Campbell,
History of California, 1914

♦ ♦ ♦ ♦ ♦

The domes housing Lick Observatory's family of telescopes grace the top of Mount Hamilton near San Jose, California. (Lick Observatory, Regents of the University of California)

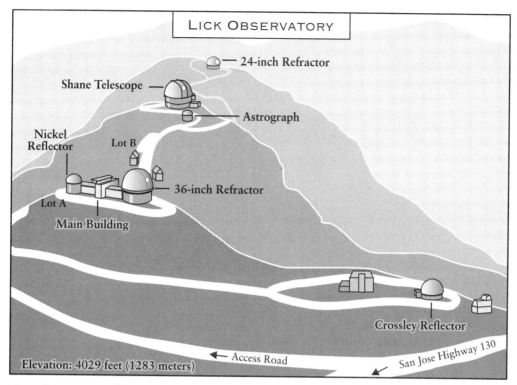

Map of Mount Hamilton (Lick Observatory, Regents of the University of California; redrawn by Jeremy Eagle)

Winding your way up the narrow, curvy two-lane road to the top of Mount Hamilton near San Jose, it's easy to see why this location became a place for seeing. Vistas open up around every bend—south to San Jose and Monterey and north toward San Francisco. Later, near the top of the mountain, you can see east to the Sierra Nevada and to Half Dome in Yosemite National Park, and your view can travel west to the Pacific Ocean. Out over the valleys a layer of fog often blankets the scene, but on Mount Hamilton the view is clear. It's a good spot for looking at stars.

And that is exactly what it has become: On this mountaintop the University of California has built a series of astronomical observatory domes to house the collection of huge telescopes that is known as Lick Observatory. But the story of how the observatory got started and how it happened to wind up on this mountain has nearly as many remarkable turns as the road you take to get there.

It began in the spring of 1873, when James Lick, at the age of 77, began to think about what he could do to perpetuate his name. He had made a fortune in San Francisco real estate, had never married, and had no heirs. Although he had never had any personal interest in astronomy, he thought he would like to provide the funds to build a telescope that would be, as he described it, "larger and more powerful than any existing."

Born in Fredericksburg, Pennsylvania on August 25, 1796, Lick had begun his career as a piano and organ maker. He had traveled to South America, where he had built a good business, and in 1848, just before the beginning of the California gold rush, he arrived in San Francisco with $30,000 in gold doubloons packed in an iron chest. He made astute real estate investments and netted a fortune from property he bought in San Francisco, in the Santa Clara Valley near San Jose, and in southern California.

By 1874, he had a sum amounting to $3 million, which he turned over to a board of trustees, setting aside $700,000 "for the purpose of purchasing land, and constructing and putting up on such land . . . a powerful telescope, superior to and more powerful than any telescope yet made . . ." He wanted it to become the property of the University of California, under the direction of the university's board of regents, and he wanted it named the Lick Astronomical Department of the University of California.

The gift was a great windfall for the six-year-old university located in Berkeley, across the bay from San Francisco. But, while Lick was clearly knowledgeable about real estate, pianos, and organs, he didn't know a lot about observing stars, and his first thought was to place the observatory in downtown San Francisco. Most observatories at the time were commonly located near universities or cultural centers for the convenience of the astronomers, who didn't relish the idea of spending lonely nights on a cold mountainside. But the idea of bogging down a great observatory on the foggy San Francisco Peninsula seemed unworkable to UC professor George Davidson, whom Lick consulted about the project. So Davidson convinced Lick that a mountain location would work better—having more clear nights and a stabler atmosphere. Numerous locations came under consideration, including mountain areas of Wyoming and Colorado, as well as the Sierra Nevada range along the border between California and Nevada. Lick favored locations on Mount St. Helens and near Lake Tahoe. But an associate of Lick's, Thomas Fraser, rode on horseback to the peak of Mount Hamilton in 1875 and as a result suggested that it might work very well. At 4,213 feet, it was high enough to gain a clear view and stable atmosphere, and yet it was close to San Jose and not far from Berkeley.

Lick was convinced, but he would agree only on the condition that Santa Clara County build a good road to the summit. The county cooperated and the road, built entirely by hand, was completed in the fall of 1876. The state and federal governments granted most of the 2,500 acres originally set aside for the observatory.

The telescope planned for the new observatory would in fact be the largest refractor telescope in the world, when it was finished, and the father-son team of Alvan Clark (1804–87) and Alvan Graham Clark (1832–97) was hired to grind the lens. It was a great coup for the Massachusetts pair; at that time the premier lens makers in the world were all English.

Construction of the observatory was a long and arduous process because every stone and timber, as well as the telescope itself, had to be hauled by horse and wagon up the narrow dirt road. But by the summer of 1888, all the work was finally completed, and the world's first permanently staffed astronomical observatory, equipped with the most powerful telescope yet attempted, was ready to set to work.

Unfortunately, neither the elder lens maker Alvan Clark, nor benefactor James Lick lived to see completion of the bold project. Clark had died the previous year, and Lick had died 12 years before, in 1876, at the age of 80.

Over the years, numerous important observations have been made at the Lick Observatory, among them two major discoveries by American astronomer Edward Emerson Barnard, who joined the Lick Observatory staff in 1890 after completing his studies at Vanderbilt University.

The 36-inch refractor was the largest refractor in the world when it was installed. (Lick Observatory, Regents of the University of California)

The Shane 120-inch telescope, the biggest telescope now at Lick Observatory (Lick Observatory, Regents of the University of California)

In 1892, while studying a nova that appeared in the constellation Auriga, he was the first to notice that it had given off a puff of gaseous matter, providing a clear indication for the first time that a nova might involve some sort of explosion. That same year, Barnard also discovered a fifth satellite (or moon) of Jupiter. No new satellites of Jupiter had been discovered since Galileo saw the first four nearly three centuries before. Closer to Jupiter and far smaller than the others, it is often called Barnard's satellite. The telescope's excellent ability to resolve fine detail has made it ideal for photographic surveys of the moon and planets, for which it was used by the observatory's first director, E. S. Holden. The 36-inch refractor was also a key instrument used in the study of double stars, thousands of pairs of which have been discovered using this telescope.

Lick Observatory retained its status of having the largest refractor in the world only until 1897, when its 36-inch refractor was surpassed by the 40-inch refractor installed at the University of Chicago's Yerkes Observatory. The Yerkes lens was also made by young Alvan Clark, who completed supervision of its construction just before his death. But no larger refractor than that has even been made. All larger telescopes have been reflectors, which instead of lenses use very large mirrors. They are easier to build and involve fewer mechanical difficulties.

Long committed to remaining open to visitors, Lick Observatory encourages public interest in developing a more complete vision of the universe. And it remains a premier astronomical research laboratory as Lick hoped, "made useful in promoting science."

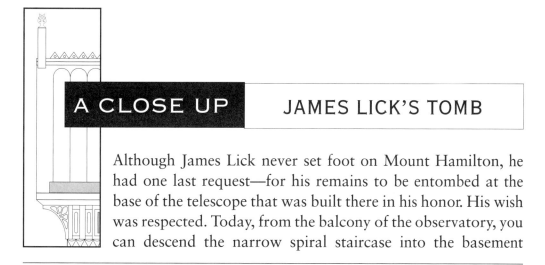

A CLOSE UP   JAMES LICK'S TOMB

Although James Lick never set foot on Mount Hamilton, he had one last request—for his remains to be entombed at the base of the telescope that was built there in his honor. His wish was respected. Today, from the balcony of the observatory, you can descend the narrow spiral staircase into the basement

The eye end of the 36-inch refractor showing the automatic camera mounted for direct photogra-phy. (Lick Observatory, Regents of the University of California)

below the observatory floor to view the tomb, which is brightly lit and often decorated with flowers. In large gold letters the inscription reads: "Here lies the body of James Lick."

Above the tomb, the floor of the observatory raises and lowers to place the observing astronomer at the eyepiece of the telescope, which changes position according to the quadrant of the sky being observed. So the floor may rest directly above the tomb, or at the balcony level, or in between.

When the observatory floor is low, the balcony can be reached using a steep, narrow metal stairway. The walls of the observatory are inlaid with wood paneling and, while the equipment has been updated—an automatic camera has been attached to the telescope's eyepiece for direct photography—the setting exudes a Victorian charm and a sense of richness that seems fitting for the study of the stars and planets, and one feels sure that James Lick would have approved.

PRESERVING IT FOR THE FUTURE

♦

The University of California is proud of Lick Observatory's history and has maintained Lick's tomb and as much else as possible as it was when it was first built. But because this is a working observatory, and astronomers still use the famous 36-inch refractor telescope for research, some changes have been made to keep up with the times. The Lick Observatory is now managed by the University of California campus at Santa Cruz.

EXPLORING ♦ FURTHER

Books about James Lick and the Lick Observatory

Lick, Rosemary. *The Generous Miser: The Story of James Lick of California.* A foreword by Richard H. Dillon and introd. by C. D. Shane. Los Angeles: Ward Ritchie Press, 1967.

Osterbrock, Donald E., John R. Gustafson, and W. J. Shiloh Unruh. *Eye on the Sky: Lick Observatory's First Century.* Berkeley: University of California Press, 1988.

Related Places

Maria Mitchell Birthplace House
1 Vestal Street
Nantucket, MA 02554
(508) 228-2896 or (508) 228-9198

Birthplace of the first professional woman astronomer, Maria Mitchell (1818–89). Highlights include the brass telescope used by Mitchell to discover Comet Mitchell 1847 VI in 1847, the only roofwalk in Nantucket that is open to the public, and special hands-on activities.

Mount Wilson Observatory
Mount Wilson Institute
P.O. Box 60947
Pasadena, CA 91116
(818) 793-3100 (voice) (818) 793-4570 (fax)
http://www.mtwilson.edu.

Founded in 1904 for solar research, this observatory in the San Gabriel Mountains of southern California is also open to public tours. A virtual tour is available at the Internet site as well.

Thomas Edison's
Menlo Park Laboratory

BIRTHPLACE OF THE INCANDESCENT LIGHT BULB
Dearborn, Michigan

AT A GLANCE

Built: Originally, 1876, by Thomas A. Edison;
moved to Dearborn, Michigan by Henry Ford and opened to the public in 1929

Laboratory of Thomas Alva Edison, 1877–1886

Originally located in Menlo Park, New Jersey, this was the first industrial
laboratory, the place where Thomas Alva Edison perfected the incandescent
light bulb and did his most famous work.

Address:
Henry Ford Museum
P.O. Box 1970
20900 Oakwood Boulevard
Dearborn, MI 48121-1620
(800) 835-5237 or (313) 271-1620
http://www.hfmgv.org

Creator of more than a thousand inventions, many historians consider Thomas Alva Edison the most prolific inventor of all time. He spent the most productive decade of his life working with his assistants in this laboratory, relocated from its original location in Menlo Park, New Jersey.

Genius is one percent inspiration and 99 percent perspiration. . . . Everything comes to him who hustles while he waits.

—Thomas Alva Edison

♦ ♦ ♦ ♦ ♦

Thomas Alva Edison's Menlo Park laboratory   (From the Collections of Henry Ford Museum and Greenfield Village)

Thirty years after Thomas Edison's death, science writer L. Sprague de Camp wrote, "With 1,093 patents to his name, [Edison] was the most productive inventor in the history of the United States, and possibly the most productive in the history of the human race." De Camp's admiration was shared by many, and with good reason. To Thomas Alva Edison (1847–1931), we owe the microphone (1877), the phonograph (1878), and the incandescent light bulb (1879), to name only a few of his many inventions. And it was Edison's fertile mind that also figured out how to distribute electrical power to ordinary households, resulting in the installation of the world's first central electric power plant in New York City (1881–82).

Today, as the lights burn from the windows of Edison's reconstructed laboratory building at Greenfield Village, in Dearborn, Michigan, the reflected gleam on the bronze statue of Edison outside makes the figure look so real, you can almost imagine him sitting there thinking up his next innovation. Inside the building, upstairs, his laboratory appears only momentarily quiet, as if his assistants had just left for the day to eat their supper in Sarah Jordan's nearby boardinghouse, where many of them lived. Wooden chairs are pushed away from worktables and the burnished brown wood of the laboratory floors gleams warmly in the light from the gas jets that illuminate the room. The intricate machinery of inventions in the making fills up tables ranged along one wall, while portable step-stairs allow access to neatly labeled jars lining the shelves of the long, narrow room. More mysterious gadgets and gizmos are stored beneath some of the tables. Perhaps any moment Edison will stroll in, talking excitedly with a small group of workers who after dinner will be putting in more of their many hours of overtime. For this is a place built for production, above all, and teamwork, the tables forming one long, continuous bench, as if each project flowed into the next.

The laboratory, along with the office downstairs and the nearby boardinghouse, were originally located in Menlo Park, New Jersey. But one of Edison's rare friendships resulted in relocation of the entire complex to its current midwestern home, more than 600 miles away. Although many people admired Edison, his temperament and preoccupation with his work didn't permit him many close friends, but in his later years he developed a bond of mutual admiration with inventor-industrialist Henry Ford. And when Ford began putting together a museum in Dearborn, Michigan to honor American

Sarah Jordan's boardinghouse in Greenfield Village (From the Collections of Henry Ford Museum and Greenfield Village)

invention, he moved the abandoned Menlo Park laboratory to Dearborn in the late 1920s, where it remains on view today.

But, of course, Edison's story was not always the famous tale of success that it later became. Born in 1847 in Milan, Ohio, Edison had a harsh childhood, marked by thrashings from both his mother and father when he got out of line. And he probably did get out of line often, with his laboratory in the basement, where he brewed up chemical concoctions and experiments that his father claimed would blow up the entire household.

Thomas's parents withdrew him from school after only three months of formal education, following his teacher's calling him "addled" and unadaptable. (Looking back, historians think that Edison very likely had a learning disability, possibly dyslexia.) Angry at the lack of understanding he received at school, from that time on, his mother taught him at home. At the age of 12 he went to work on the train that ran between Port Huron, Michigan and Detroit. He sold candy and newspapers, one of which, enterprising as he was, he wrote and published himself. He also haunted the bookshelves of the Detroit Free Library, where he read everything he could get his hands on.

Eventually, through some quick talking, young Edison was allowed to move his laboratory to the baggage car of the train. And there he continued his experiments until what his father had always feared finally happened—he set the place on fire—at which time, as the story goes, his train-riding days came abruptly to an end. Years later, when Henry Ford began bringing historic Michigan structures to Dearborn to create the place he called Greenfield Village, the 1858 Smith's Creek Grand Trunk Railroad Depot numbered among them. It's the very depot where "Little Al" Edison was unceremoniously booted off the train, along with all his chemistry gear.

Edison's interest extended beyond chemistry, though. As a boy, he had begun learning Morse code and made his own telegraph set to practice on. So when the Civil War (1861–65) produced a demand for telegraphers, he was ready to step in, and in 1863, at the age of 16, he began working at various Western Union outposts. Always fond of fiddling with contraptions, by the time he was 21 he had succeeded in inventing a recording stock ticker, which proved very useful in the postwar economic boom. Thomas Edison's career as an inventor had begun.

He set up his first shop in Newark, New Jersey to produce the stock tickers himself, hiring skilled craftsmen to help, several of whom worked at his side for most of their lives. A team began to form, and Edison continued his inventing, turning now to refining the telegraph, which could either send or receive, but not both at once. By 1874 he had come up with a method for sending four messages at the same time, two going out and two coming in.

By the spring of 1876, when Edison was nearly 30, he decided he needed a place to retreat to, where he and his hand-picked workmen could concentrate more fully on his work. He bought a plot of land in Menlo Park, New Jersey, about 25 miles southwest of New York City. There he established what historian Matthew Josephson has called "the first industrial research laboratory in America, or in the world, and in itself one of the most remarkable of Edison's inventions."

He gathered about himself a group of associates, all men, whose lives he ran with a firm hand. To work with Edison was both a privilege and an intense, demanding experience. He expected, and got, complete loyalty and tireless work. Edison was reputed to have warned applicants with only the slightest bit of exaggeration, "Well, we don't pay anything, and we work all the time." An average work week was 10 hours a day and six days long, with many much longer days when a push was on. John Ott, who had been with Edison since he was first hired by the inventor to help manufacture stock tickers, once remarked, "My children grew up without knowing their father. When I did get home at night, which was seldom, they were in bed."

Not surprisingly, many of the most loyal men were unmarried and lived in the tidy rooms of the nearby boardinghouse run by the industrious Sarah Jordan. Her reconstructed boardinghouse is among the structures on view in Greenfield Village. There you can see the small, Spartan rooms they lived in, with quilt-covered beds, filmy curtains, and plain, serviceable furniture. Furnishings didn't matter much to these men, though, as long as everything was neat and clean, as it always was at Sarah Jordan's. In fact, the men spent little time in these rooms—when they weren't at the "invention factory," they played billiards or drank a pint of beer at the corner tavern.

Edison's great fascination was with technologies for communication, and this underlying theme infuses nearly all of his inventions and theories. He was methodically developmental, working from his own often sketchy drawings and moving forward step by step, frequently using analogies to stake out the progressive presentation of problem and solution. However, the inventions from this laboratory, like all Edison inventions, were the labor of not just his own but many hands.

By the summer of 1876, Alexander Graham Bell began to win acclaim throughout the country for his demonstrations of his new invention the telephone. In a complex drama of industrial intrigue, Western Union hired Edison to invent a better telephone, which ended in a long and windy patent dispute. But what Edison did come up with was an improved transmitter that made Bell's invention work much better. Edison's small disk of carbon, which he used to vary the electrical current, worked far better than Bell's original system.

On July 18, 1877, Edison came up with a very primitive phonograph, followed about four months later with a second, more effective one. It had a revolving cylinder, which the user turned with a hand crank. A piece of tin foil wrapped around the cylinder provided the surface to be recorded on, which Edison did, making his famous recording of the nursery rhyme "Mary Had a Little Lamb." Elated, he packed up his machine and set out for Washington, D.C., where he showed it off to Joseph Henry (founder of what became the Smithsonian Institution), as well as members of Congress and President Rutherford B. Hayes. The public and the press were caught up in the excitement. To many it seemed Thomas Edison had somehow captured time itself and sealed it up to be played back whenever the hearer wished. Due to the tremendous excitement generated by the phonograph, Edison acquired the title "Wizard of Menlo Park"—*wizardly* seemed the only word to describe the phonograph at the time. Ironically, though, Edison himself could never completely appreciate the sounds of his invention. He had suffered a progressive hearing disorder that had made him nearly deaf since childhood, and to listen to the sounds and music played on his invention was difficult for him. Sometimes, to get a better sense of the sound, he would bite down on the flared horn of the speaker so vibrations of sound would be carried by the bones of his head.

In 1878, on a trip to Wyoming to view an eclipse of the sun, Edison got into a conversation with a traveling companion that focused his mind on a new project. It was a project that would gain him even greater fame: producing an electric light that people could really use. Some 70 years earlier, physicist Humphrey Davy in England had demonstrated an electric arc light, but it wasn't practical. Still, the possibilities of using electricity to produce lighting had intrigued scientists and inventors ever since. British scientist Joseph W. Swan had worked on the idea in 1850, noting that carbon glowed when excited by electrical energy. And Edison also came across this fact as he explored the possibilities of an incandescent lamp—a bulb that emits a glow caused by electrical heating.

Edison became totally focused on his project, determined to produce a marketable product and produce it fast. Enormously famous by now, he had not only his own demands to drive him—when word got out on what he was working on, the expectations of a waiting public drove him as well. He and his assistants worked day after day, 20 hours a day, for weeks. Finally they produced an incandescent lamp that glowed a faint red and could burn for 40 hours. The date was October 19, 1879.

As Henry Ford and Frank Jehl look on, Thomas Edison reenacts the invention of the incandescent light bulb for the Golden Jubilee celebration in 1929. (From the Collections of Henry Ford Museum and Greenfield Village)

The invention would only be useful, however, if electricity to burn the lamp could be delivered to consumers. Over the next three years, using a trial-and-error process, Edison worked through a complicated series of hurdles to get a basic system going. Because natural gas was already piped to houses in the big cities, providing gas light, electrical wires could be threaded along the pipes. So the main challenge was to build a reliable system of dynamos to produce the electricity cheaply and steadily. But that wasn't all. He also had to find a way to seal the bulbs, devise screw-in sockets, and build safety devices, not to mention light switches, meters to measure electrical flow, and a throng of other mechanisms. But he and his loyal team put the whole system together within 15 months. It was one of Edison's greatest achievements.

In a low-rent district in lower Manhattan, he bought some property and built a power station on Pearl Street. And on September 4, 1882, with only 85 subscribers and one dynamo, Edison turned on the juice. Two years later, the number of subscribers had expanded to 508 in New York, with demand growing, and power stations springing up in other cities as well. A revolutionary way of living had begun, with a network laid that would one day bloom into today's households equipped not only with electrical lighting but also with microwave ovens, food processors, washing machines, clothes dryers, computers, TVs, VCRs, and CD players, power saws, electric drills, and screwdrivers, and more—all run on electricity.

In 1884, Edison's wife Mary Stillwell, whom he had married when she was only 16, contracted typhoid fever and died. Edison was 37. Two years later he was remarried, to Mina Miller of Akron, Ohio. In 1886, Edison closed down the laboratory at Menlo Park and built a new laboratory and home in West Orange, New Jersey (which are now both part of a National Historic Site) and moved there with his new wife. Here he worked more slowly, still relying on his top team of 50 or more assistants. During the 1890s a flow of useful inventions came out of the West Orange lab, including a fluoroscope, a dictating machine, a mimeograph, and the strip kinetograph—the first motion picture camera.

In the latter part of his career, preoccupied with the business end of his inventions, Edison didn't always make brilliant calls, however. In the controversy over whether electrical current should be direct (DC) or alternating (AC), he favored DC. But by the mid-1880s, George Westinghouse (with the help of former Edison associate Nikola Tesla) had won the battle for AC by demonstrating that a practical transformer could make it work well and over long distances at higher voltages. Also, in 1883 Edison had pioneered the vacuum tube eventually used in radios when he found a method for passing electricity from a filament to a plate of metal inside an incandescent light globe. He patented this "Edison effect"; yet he never had any use for the field of electronics, which is based on his discovery.

Some of Edison's other inventions failed to catch on—a storage battery for powering an electric car was outpaced by Detroit's gasoline-powered engines (for the time being, at least), and an inexpensive prefabricated concrete house he designed found few takers. He lost almost $2 million trying to extract iron from low-grade ore in New Jersey. But his improvements on the phonograph and his work with motion picture film put him at the cutting edge in those fields and contributed to his fame.

Running out of inventiveness and energy in his later years, he still enjoyed basking in his fame. And in the company of his friend Henry Ford he would often take long "camping" trips, which were actually long and elaborate and well-publicized excursions made in Ford cars and trucks. On October 21, 1929, the Golden Jubilee—50th anniversary—of the birth of the incandescent light bulb, Henry Ford opened his showcase in Dearborn. Thomas Edison was there for the triumphant celebration and reenactment, as Ford stood proudly by.

Afflicted with kidney disease, uremia, diabetes, and a gastric ulcer, Thomas Edison died October 18, 1931, just short of the 52nd anniversary of the moment that changed the world.

A CLOSE UP — THE LABORATORY PIPE ORGAN

Edison's Menlo Park laboratory was not a place where work was a dull task. Recognizing the positive effect of music on the creative mind, Edison installed a pipe organ for his workers, which visitors can still see at the far end of the room. So massive that it dominated that end of the laboratory, the organ was given to Edison by its manufacturer, Theodore Roosevelt's cousin Hilbourne Roosevelt.

The pipe organ at the back of the lab was an important ingredient in the creative process. (From the Collections of Henry Ford Museum and Greenfield Village)

Edison loved to fill the lab with sonorous chords from this magnificent instrument, improvising tunes of his own composition. In the evenings, the men of the shop would gather round the organ and sing popular songs of the day, ranging from evangelistic hymns to bawdy ditties.

PRESERVING IT FOR THE FUTURE

In the 1920s Henry Ford conceived his idea for a museum to celebrate American innovation and the spirit of American progress, and he began collecting antiques and buildings—most dating from the 19th century and gleaned from the surrounding Michigan countryside. But the centerpiece was to be the early laboratory of his friend Thomas Alva Edison.

When he and Edison visited the site in 1928, however, most of the buildings had been moved or demolished. Undaunted, Ford collected what remnants he could from the site—boards, bricks, debris, and even the red New Jersey clay beneath the foundation—and shipped them from New Jersey to Michigan. There he had the laboratory reconstructed based on photographs. He purchased some of Edison's old tools from a blacksmith who had bought them a few years back when Edison had cast them off.

Many luminaries attended the opening of the Ford Museum and Greenfield Village and Golden Jubilee extravaganza that Ford organized to celebrate Edison's invention of the incandescent light bulb. Among them were physicist Marie Curie; industrialist John D. Rockefeller, Jr.; steel magnate Charles Schwab; inventor George Eastman; and humorist Will Rogers.

Today the museum houses a collection of more than one million three-dimensional artifacts, as well as an enormous archive, focusing on communication, transportation, agricultural and industrial production, and domestic life. The adjacent 81-acre Greenfield Village contains 80 historic structures, including Thomas Edison's Menlo Park Laboratory.

EXPLORING ◆ FURTHER

Books about Thomas Edison and Invention

Adair, Gene. *Thomas Alva Edison: Inventing the Electric Age*. Oxford Portraits in Science series. New York: Oxford University Press, 1996.

Amrine, Michael, et al. *Those Inventive Americans*. Washington, D.C.: National Geographic Society, 1971.

Baldwin, Neil. *Edison: Inventing the Century*. New York: Hyperion, 1995.

Billington, David P. "Edison and the Network for Light," in *The Innovators: The Engineering Pioneers Who Made America Modern*. New York: John Wiley & Sons, Inc., 1996.

Cousins, Margaret. *Story of Thomas Alva Edison*. Landmark Books, Vol. 8. New York: Random House, 1981.

Parker, Steve. *Thomas Edison and Electricity*. Science Discovery Series. New York: Chelsea House, 1995.

Related Places

Edison National Historic Site
Main Street and Lakeside Avenue
West Orange, NJ 07052
(973) 736-5050

In 1886, at the age of 40, Edison moved his laboratory to West Orange, New Jersey, the "invention factory" where, for the next 44 years he employed some 50 assistants. Here he invented the kinetograph, which was the first motion picture camera. Visitors can see the laboratories and view originals of many early Edison inventions, as well as a replica of the world's first motion picture studio, Black Maria. Edison's company made some 1,700 silent movies, including *The Great Train Robbery*, which visitors can view here. About a mile away, also open to the public, is Glenmont, the opulent Victorian mansion where Edison lived with his second wife, Mina. Edison's "thought bench," as he called the desk where he thought up many of his inventions, can be viewed in the second-floor sitting room. The site is managed by the National Park Service.

Thomas A. Edison Winter Home

2350 McGregor Blvd.
Fort Myers, FL 33901
(813) 334-3614

The winter retreat where, for decades, Edison spent his working vacations, surrounded by profuse gardens. Edison first visited in 1885 and he liked the small, slow-paced town so much that he decided to winter there regularly. He built his winter home in 1886. But his way to relax was to work incessantly. So here, too, visitors can view another of his laboratories crammed with a forest of test tubes and other apparatus. Here he experimented with making rubber from the sap of the goldenrod flower, raised bees to provide wax for his phonograph records, and worked on his light bulbs and many other inventions. On display are numerous Edison inventions, including a ticker tape, the first phonograph, the oldest electric lamp, and a talking doll, as well as his nickel-alkaline storage battery.

Thomas Edison Birthplace Museum

9 Edison Drive
Milan, Ohio 44846
(419) 499-2135

The house, built in 1841, where Edison was born in 1847 and lived until he was seven years old, with exhibits about Edison's life, featuring several of his inventions. The house became a registered national historic landmark in 1965.

Luther Burbank Home and Gardens

HOME AND LABORATORY OF "PLANT WIZARD" LUTHER BURBANK
Santa Rosa, California

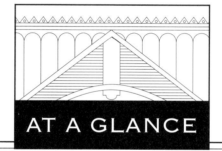

AT A GLANCE

Built: 1883

Home of horticulturist Luther Burbank, 1884–1906

Modified Greek Revival house, where Luther Burbank lived for 22 years, and 1.6 acres of the 4-acre garden that served him as an outdoor laboratory for his horticultural experiments.

Address:
Luther Burbank Home and Gardens
P.O. Box 1678
Santa Rosa and Sonoma Avenues
Santa Rosa, California 95402
(707) 524-5445

Every child should have mud pies, grasshoppers, water-bugs, tadpoles, frogs and mud-turtles, elderberries, wild strawberries, acorns, chestnuts, trees to climb, brooks to wade in, waterlillies, woodchucks, bats, bees, butterflies, various animals to pet, hayfields, pine-cones, rocks to roll, sand, snakes, huckleberries and hornets; and any child who has been deprived of these has been deprived of the best part of his education.

—Luther Burbank

♦ ♦ ♦ ♦ ♦

The enthusiasm Luther Burbank (1849–1926) had for the natural world began early in life and by the time he was 50, he was known far and wide as the "plant wizard." He seemed to know how to get plants to grow in ways they never had before—creating, among many other successes, such surprising varieties as the white blackberry, the pitless prune, and the spineless cactus. People in Santa Rosa, California were proud of him—he had singlehandedly put their little town on the map. But while Burbank's work seemed like wizardry to his neighbors, it was also highly respected among the community of scientists early in the 20th century.

The field of genetics and hereditary science was still in its infancy when Burbank began his work. Darwin completed his voyage aboard the HMS *Beagle* only 13 years before Burbank's birth, and Darwin's results, published in *On the Origin of the Species* in 1859, came out when young Luther was only ten. And the work of the great experimenter Gregor Mendel—a monk who made important early discoveries about heredity—was completely

Luther Burbank   (Luther Burbank Home and Gardens)

unknown until it was rediscovered independently by three separate investigators in 1900.

Of course, farmers and animal breeders had been using many of the principles of natural selection and hybridization for centuries—producing new varieties or species of plants or animals through cross-breeding. But before Darwin and Mendel no one had combined careful observations with accurate record keeping and the ability to formulate theories. So, while Burbank read Darwin's work on natural selection and was greatly impressed by it, he did not have the advantage of seeing Mendel's work on the hybridization of peas until the beginning of the 20th century, when Burbank had already done much of his work.

Luther Burbank started out, in fact, as a gardener, but by the time he was 20 he knew he wanted to be a plant breeder. Born March 7, 1849 in Lancaster, Massachusetts, he went to work at age 15 for the Ames Plow Company in Worcester, Massachusetts, which he left in 1867 to attend the Lancaster Academy for two years. That's where he first encountered the writings of Darwin.

Burbank's father died in 1870, leaving a small inheritance with which Burbank bought a plot of land. There he went into business at 21 as a truck farmer, producing vegetables for sale in a nearby town. But he also began experimenting. He saved the seeds from the best and sturdiest plants in his gardens, replanting and nurturing them through several generations. During this period he developed the "Burbank Russett" potato, which was a larger, whiter potato than the small, reddish potato that was common at the time.

Burbank sold the rights to his potato seed for $150, enough to buy a train ticket and get himself settled in California, where he knew he would be able to grow plants year-round. He arrived in Santa Rosa in 1875 and bought a small fruit and vegetable farm the following year. There, he continued his experimentation.

By combining practical experience and Darwinian evolutionary theory, Burbank began to get results. He used cross-breeding, hybridization, and grafting to develop new varieties that had superior qualities in color, taste, resilience, or other desirable characteristics. He produced new berry varieties by 1878, prunes without seeds in 1879, blackberry bushes that had no thorns in 1880. By 1882, he had new varieties of the gladiolus flower, and by 1884,

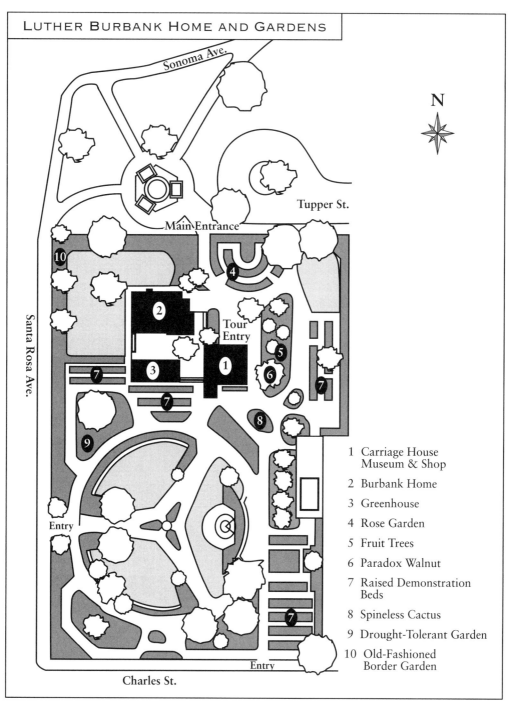

LUTHER BURBANK HOME AND GARDENS

Sonoma Ave.

N

Tupper St.

Main Entrance

Santa Rosa Ave.

Tour Entry

10

4

2

3

7

7

1

5

6

7

8

9

Entry

Entry

Charles St.

1 Carriage House Museum & Shop

2 Burbank Home

3 Greenhouse

4 Rose Garden

5 Fruit Trees

6 Paradox Walnut

7 Raised Demonstration Beds

8 Spineless Cactus

9 Drought-Tolerant Garden

10 Old-Fashioned Border Garden

The grounds of the Luther Burbank Home and Gardens   (Luther Burbank Home and Gardens; redrawn by Jeremy Eagle) he had developed improved varieties of the quince, persimmon, loquat, chestnut, pear, and peach.

He gave up truck farming the following year, turning exclusively to experimentation, for which he had enormous stamina, an intuitive knack for preserving traits that best suited a plant for survival, and a willingness to pursue improvements doggedly through generation after generation of plants. He sold his results to local nurseries at first, and from 1893 to 1901 he distributed his *New Creations* catalog, through which he sold seeds and root stock.

Burbank's technique involved finding a specimen that had a characteristic he wanted to perpetuate and crossing it with one that had other characteristics that would make it strong, healthy, and in other ways desirable. To obtain a white blackberry, for instance, he crossed a Southern bramble, having a pale yellow fruit, with a popular blackberry. The result was an offspring that had the color of one parent and the edible characteristics of the other. He often imported unusual varieties—which were frequently considered useless—to cross them with native plants, and he had feelers out all over the world for news of specimens that might interest him. In the case of the Shasta daisy, one of his most enduring creations, he combined wild American species with European and Japanese plants to produce the unique hybrid that is still popular today.

But most of all the secret of his success was the vast scale of his experiments, matched by no other experimenter. He has been called the Henry Ford of hybridizing, using mass production, with thousands of seedlings, to produce one single improved variety. With the sheer numbers he was working with, the chances of finding a useful mutation, or "sport," were vastly increased. And this increase in possibilities was necessary, since Burbank often was trying to perpetuate five or six qualities, out of which normally only three or four would show up in any individual plant.

Burbank also developed mass methods for grafting the branches of one tree onto the root stock of another. He had hundreds of stock trees on his place, and he would graft each one with dozens of different hybrid plums. One tree could have an amazing variety of foliage and fruit—the objective being to find the fruit that best achieved his goals, without having to wait the years it would take for a seedling to grow and produce fruit. Moreover,

Burbank developed many different varieties of flowers, including the Shasta daisy. (Luther Burbank Home and Gardens)

the resilient characteristics of the root stock provided a strong start that varieties having less well adapted root systems might never otherwise get.

One of the scientists who rediscovered Mendel, the Dutch botanist Hugo De Vries, visited Burbank a few years after finding Mendel's meticulously documented work. Like Burbank, Mendel had carefully controlled the parentage of his plants, pollinating each plant by hand. But he also kept painstaking records and did careful statistical analysis of his results—providing a sound basis from which to develop a scientific explanation for hybridization. De Vries was quick to praise Burbank for his "special gift of judgment, in which he excels all his contemporaries"—high praise from a man who had studied with expert plant breeders all over Europe.

But, while Burbank's methodology got results, it was, in the opinion of De Vries, impossible to imitate because the process that took place in Burbank's mind was too complicated "for simple imitation." Moreover, Burbank was never systematic about gathering scientific data. In 1905, the Carnegie Institution awarded Burbank $10,000 a year (a very sizable sum at the time) and an assistant to record his work for genetic research. But Burbank couldn't be bothered with recording details he could keep in his mind, and the Carnegie Institution withdrew its support in 1910.

From 1905 to 1911, Burbank taught a course in plant breeding at Stanford University, during which time he continued his experimentation. He also directed the writing of a 12-volume series entitled *Luther Burbank: His Methods and Their Practical Application*, completed in 1915.

Luther Burbank died in 1926 at the age of 77, and he was buried beneath a cedar of Lebanon tree he had planted in front of his house in 1893. In 1989, the nearly 100-year-old tree had to be removed because of root disease, but some of its wood has been preserved as sculptures and benches. In 1909 the California legislature established Arbor Day as a state holiday in Burbank's honor, and trees are still planted annually in his memory on Arbor Day in California.

Visitors to the Luther Burbank Home and Gardens in Santa Rosa can tour his house, where he lived from 1884 to 1906. The house contains memorabilia and the original furnishings, and his adjacent Victorian greenhouse, built by Burbank in 1889, offers exhibits as well as a replica of his office and a display of many of his tools. But the real focus of interest is his gardens—originally covering 4 acres—which, though now trimmed down to 1.6 acres, are filled with many fascinating examples of his hybrids. Here visitors are encouraged to enjoy the butterflies, birds, and bees; to breathe in the aroma of Burbank's flowers; to listen and touch; and to see the results of this tireless experimenter's work.

The rose garden features the Burbank rose, which the horticulturist developed for its pleasant fragrance. At least 20 others on display are known for their special fragrances, with names such as Allspice, Lemon Spice, Double Delight, and Sweet Surrender.

Additionally, a Victorian garden features flowers popular 100 years ago, when Burbank lived here, such as sweet-smelling lilacs and roses and a bush with dark purple flowers known as a Butterfly Bush, which attracts butterflies. Other creatures also can often be observed thriving here, including birds, bugs, and lizards, exhibiting the close interdependence of plant and animal life that Rachel Carson wrote about 50–60 years later.

The Drought-Tolerant Garden displays plants, most of which are indigenous to California, that thrive in regions where water is scarce. Burbank worked with many of these plants, increasing their flower size or improving their color, introducing them to gardeners through his seed catalog. Drought-tolerant plants are of particular interest in California, where water conservation is a high priority.

Burbank's work with fruit trees—he introduced over 200 varieties of fruits and nuts, including more than 100 plums and prunes—is represented by a tiny orchard near the front of the house, as well as other trees around the garden's perimeter and espaliered (trained to grow flat) on the fences. His plumcot tree (a cross between a plum and an apricot), which most people thought was impossible to produce, grows in the orchard, and other exhibits of his expertise include a multiple-grafted cherry tree with four different kinds of cherries on it. Burbank's contributions in this area were extensive, including many different apples, berries, cherries, figs, grapes, nectarines, peaches, quinces, chestnuts, and walnuts.

The demonstration beds near the greenhouse are a series of raised beds patterned after those used by Burbank both at this site and at his experimental farm in nearby Sebastopol. Visitors can view a variety of plants here, depending on the season, many either developed by Burbank or very similar to those he grew, including lilies, dahlias, zinnias, asters, gladioli, the famous Shasta daisy, and various vegetables and herbs. The "Burbank Russet" potato that got him started in his career is here—and visitors are encouraged to look under the straw in the bed to see the potatoes growing on the plant stems.

A CLOSE UP   THE SPINELESS CACTUS

Everyone knows that cactuses (or cacti, as they are more properly called in the plural) are prickly. So how did Luther Burbank come up with one that had no spines—his "spineless cactus"?

His system was, as always, practical and methodical. He collected varieties of cactus from countries all over the world —Mexico, South America, and other regions, often from countries to which they had been exported from the Western Hemisphere—in hope of finding one with the characteristics he wanted. Finally he found one that had no spines on its stalks, although it had spines on its leaves. Another had no spines on its leaves, although it had spiny stalks.

After an extensive series of crossings, he hybridized these characteristics into a single plant. The result was a cactus you could comfortably rub against your cheek—as the affable man liked to demonstrate to his visitors. It was a product that Burbank believed would be popular as forage for cattle in desert areas, where it would grow well, would retain water, and would be easy to eat.

Spanish vaqueros and Indians had long used chopped cactus plants—their spines burned off—as a last resort to feed their cows, so Burbank's idea wasn't impossible. But Burbank's detractors pointed out that the spines on ordinary cactus kept it from being eaten at early stages of its growth, and that growing cactus as a feed crop would require irrigation. If irrigation was necessary anyway, why not grow alfalfa instead?

The spineless cactus was one of Burbank's less successful products, although visitors enjoy its novelty.   (Luther Burbank Home and Gardens)

But Burbank did find some customers, and while it never really has caught on as more than a novelty, visitors have made the spineless cactus in the Burbank gardens a great favorite.

PRESERVING IT FOR THE FUTURE

The Luther Burbank Property became a Registered National Historic Landmark in 1964, and it is also listed as a State and City Historic Landmark. In 1977 Burbank's widow, Elizabeth, gave the Burbank home property—including the home, experimental garden, and greenhouse—to the City of Santa Rosa. The Luther Burbank Home and Gardens Board and Santa Rosa city staff cooperatively oversee and maintain the property, which is open to the public, and the programs are staffed by volunteers.

Books about Luther Burbank

Dreyer, Peter. *A Gardener Touched with Genius: The Life of Luther Burbank.* New York: Coward, McCann & Geoghegan, Inc., 1975.

Quackenbush, Robert. *Here a Plant, There a Plant, Everywhere a Plant, Plant!* Englewood Cliffs, N.J.: Prentice-Hall, Inc., 1982.

Related Place

The Cloisters
Fort Tryon Park
193 Street and Fort Washington Avenue
New York, NY 10040
(212) 923-3700

This museum is housed in a building constructed from sections of several European monestaries, moved here by John D. Rockefeller. It includes three gardens that visitors can tour. One is an herb garden of plants grown and used—often medicinally—during the Middle Ages.

Wright Brothers National Memorial

BIRTHPLACE OF HUMAN FLIGHT
Kill Devil Hills near Kitty Hawk, North Carolina

AT A GLANCE

Date: 1903

Site of the first successful flight of a heavier-than-air machine

On these sand dunes Wilbur and Orville Wright made the first successful sustained powered flights in a heavier-than-air machine. Markers show where the four flights began and ended, and the site includes a reconstruction of their workshop and replica of their first plane, *Flyer.*

Mailing Address:
Wright Brothers National Memorial
c/o Cape Hatteras National Seashore
Route 1, Box 675
Manteo, NC 27954
(919) 441-7430; (919) 441-7730 (Fax)

Visiting Address:

The park is located on the Outer Banks of North Carolina in the town of Kill Devil Hills. To reach the Visitor Center, travel 15 miles northeast of Manteo, N.C. on US-158. A landing strip is also available for small aircraft.

It is possible to fly without motors,
but not without knowledge and skill.

—Wilbur Wright

♦ ♦ ♦ ♦

In 1900, Wilbur Wright was 33 and his brother Orville was 29. They'd
heard that the winds along the dunes of North Carolina's Outer Banks
were among the steadiest and most predictable in the nation. The isolated

Success at last! The first successful flight, December 17, 1903. (Courtesy National Park Service)

region known as Kill Devil Hills near the tiny town of Kitty Hawk would give them privacy while they ran their tests, they thought. And the sand dunes would make for soft landings when the inevitable crashes took place.

So the two eager inventors packed up their equipment, set off from their family home in Ohio, and headed for tiny Bodie Island, precariously perched between the vast waters of the Atlantic Ocean and North Carolina's Albemarle Sound at the mouth of the Roanoke River. There, among the winging gulls, they began a series of the most outlandish of experiments with a single daring purpose: to pilot a flying machine.

For millennia people had dreamed of flying as they watched birds soar effortlessly through air. Greek mythology told the story of Icarus, who tried to fly with wings made of feathers and wax attached to his arms. But the wax melted, or so went the story, and Icarus tumbled out of the sky to his death. Leonardo da Vinci (1453–1519), the great Italian inventor and artist of the 15th and 16th centuries, had sketched plans for flying machines. And as the 19th century drew to a close, many scientists, inventors, and hacks had tried to come up with a way to fly. Some were simply publicity mongers who had no hope of conquering the involved scientific and mechanical challenges. Others nearly succeeded.

A few pioneers began to take a new look at the graceful dips and glides of birds that soar, such as hawks, gulls, and albatrosses. Perhaps, they reasoned, humans could not power their flight with flapping wings, as birds do. But they might be able to make use of the fixed and stable positions of birds' wings when they glided. In England in the early 1800s, George Cayley (1773–1857) had begun a serious investigation into the aerodynamics of gliders. He carefully wrote up the results of his experiments and even flew a few gliders with human passengers aboard for short periods in 1809. In the United States, John J. Montgomery, a California college professor, made a series of successful flights toward the end of the 19th century. But the greatest of the glider pioneers was German aeronautical engineer Otto Lilienthal (1848–96), who established the use of the curved wing. Beginning in 1891, within a five-year span he made more than 2,000 successful, often spectacular flights aboard gliders he designed. In 1896, he died of injuries sustained in a crash. The last words he murmured were: "Sacrifices must be made."

Hundreds of others were inspired by Lilienthal's flights, among them two Dayton, Ohio bicycle mechanics who believed they could succeed where others had failed. Wilbur Wright (1867–1912) was born on April 16, 1867 near Millville, Indiana. His brother Orville (1871–1948) was born four years later in Dayton, Ohio on August 19, 1871. Their father, Milton Wright, was a bishop of the United Brethren who had somewhat progressive ideas about education and bringing up children. Although Wilbur did well in school, he became ill as a result of a sports injury during his final year in high school and never fully recovered his health. Neither brother attended college and neither married. Instead, they went into the newspaper business together in 1889. But that enterprise did poorly, and they began to find they had another talent that put them in great demand: They were very able bicycle mechanics.

Nearly everyone in the 1890s had or wanted to have a bicycle. Cyclists toured the countryside in groups. They formed cycling clubs and lobbied (successfully) for better roads at a time when horses and buggies were still the primary form of private transportation. The bicycle took America by storm. So the bicycling business was a natural for the two mechanically inclined brothers. Still in their early twenties, the two young men set up shop next door to their home in Dayton and began building bicycles in 1892. They were honest and forthright—qualities their customers appreciated—and bicycle enthusiasts from all over town soon gave them a brisk business.

But they never forgot their fascination and delight born some 14 years earlier when they were still just kids and their father had given them a toy "helicopter." It was really just some bamboo, paper, and cork powered by a wound-up rubber band, but they loved it, and they began to dream the great dream: that they could build a "flying machine." When they heard about Lilienthal's last flight in 1896, they began to think about such a machine more seriously than ever.

They began to study what other pioneers had attempted and accomplished before them, mastering the available literature and knowledge of aerodynamics. They readily recognized the three major challenges that flight pioneers still faced: lift, propulsion, and control. They knew that most of the problems with lift had already been solved—only a little refinement was needed. And as for propulsion, the 1890s had seen the birth of the gasoline-powered automobile, and the Wright brothers suspected that the gasoline engine could be pressed into service aboard an aircraft as well. Control, however, posed a serious problem, one that had already brought an end to the hopes, not to mention the lives, of more than one would-be flyer.

Yet they were confident. In 1900, Wilbur Wright wrote to French flight pioneer, Octave Chanute (1832–1910):

For some years I have been afflicted with the belief that flight is possible to man. My disease has increased in severity and I feel that it will soon cost me an increased amount of money if not my life. . . . It is possible to fly without motors, but not without knowledge or skill. This I conceive to be fortunate, for man, by reason of his greater intellect, can more reasonably hope to equal birds in knowledge, than to equal nature in the perfection of her machinery.

The Wrights approached the problem scientifically, not just by trial and error, but also by studying principles, steadily sorting out fact from fiction, and logically working through the data available. "Those who tried to study the science of aerodynamics knew not what to believe," Wilbur wrote of those early days. "Things which seemed reasonable were very often found to be untrue, and things which seemed unreasonable were sometimes true. Under this condition of affairs students were accustomed to pay little attention to things that they had not personally tested."

Pragmatic and patient, the brothers approached the problem of control and stability of the aircraft. The conventional wisdom of the day dictated the idea that stability should be inherent in the design of the aircraft, but the Wrights recognized that such a design was difficult if not impossible. Also, others had attempted to keep aircraft stable by having the pilot continuously shift weight to the right or the left while gliding through the air, but the results of this arrangement were jerky and unsatisfactory. Control by the pilot was a key to success, they believed. The question was, how? They watched flying birds. They created hundreds of small models and tested them in flight. Then one day in 1899, Wilbur happened to twist a box that was constructed much like the biplane glider models they were building, and he noticed how twisting it could warp, or bend, the surface. As a result, he and Orville finally hit upon a system that allowed them to alternately "warp" the end of each wing at will to maintain stability in flight. They began testing wing-warping on a 5-foot biplane kite. Then, gaining confidence in their design, they built a 17-foot glider with an unusual front piece that acted as an elevator.

In 1900, they made their first trip to Kitty Hawk with the 17-foot glider. There, on a stretch of beach near the Kill Devil Hills, they used the strong winds the area was known for to test their design. But they were disappointed in the lift generated by the wings and had to fly the glider mostly as a giant kite, managing the warp of the wing surfaces from controls on the ground. Wilbur did pilot the glider, but his total time in flight amounted to only 10 seconds. The brothers returned to Dayton discouraged with their overall results, yet pleased with their advances in the realm of stability and control.

The following year, the Wrights began working on the lift problem, increasing the camber, or arch, of the glider and lengthening its wingspan to 22 feet. It was the largest glider that anyone had ever attempted to fly. They returned to the sands of Kill Devil Hills, where they camped again and began a new set of experiments. But the lift problem was not solved. They had based their wing design on data gathered by Lilienthal; yet the lift was only a third the altitude predicted by Lilienthal's data. The glider pitched wildly, climbed, and stalled. The Wrights changed the glider back to the earlier camber, which gave them better control, and they finally achieved a successful glide of 335 feet. But the pilot now got unpredictable results from his work with the controls. When he raised the left wing, expecting a right turn, instead the machine would slide to the left. Finally, realizing that the data they had used for their design was inaccurate, the Wrights became deeply discouraged and considered giving up. But instead they decided to collect their own data.

To test their ideas, back in Dayton the brothers built their own small wind tunnel, trying out more than 200 different wing models, meticulously collecting data on the performance of each one before settling on the arrangement that worked best. The new design was a major breakthrough, and, more important, it also worked with the full-size model that they quickly scaled up. Their new glider sported 32-foot wings, with vertical tails to counteract the directional control problem (known as adverse yaw) that they had encountered in their previous design. They also fitted it with a combined system of rudder movements and wing warp—which the pilot controlled while lying on his stomach on the bottom wing of the biplane. From this position, the pilot could change the warp of the wing by rocking in a cradle beneath his hips. This innovation later evolved to become the movable aileron, the flap at the back of an airplane wing that can be moved to control the plane's rolling and banking movements.

In the fall of 1902, the brothers returned to North Carolina, where they made some 400 glides over the sands of Kitty Hawk, proving the new design workable. But it still had problems. Sometimes, when the pilot tried to raise the lowered wing to come out of a turn, the machine still slid sideways toward the wing and crashed to the ground.

Yet they weren't discouraged. Orville suggested that a moveable tail might help correct this flaw, and Wilbur thought of linking the tail movement to the warping mechanism. By working both together, the plane could be stabilized and flown smoothly. Although the brothers often argued and debated their ideas, they were a strong team, rarely flustered by their disagreements and always focused on solving the problem at hand. This time they had worked out an important conceptual difference between a boat navigating in water and an airborne craft: a plane, unlike a boat, turns by rolling. Another 600 flights satisfied the brothers that they were on the home stretch: They had a flyable plane.

But this was still just gliding—not really flying. The next step was to power the aircraft. To solve the power problem, the brothers turned to the newly perfected internal combustion engine, which used gasoline as fuel. They designed their own four-cylinder model that was small and light enough to be fitted on the airplane; it also could give enough power for its size to operate the plane while carrying the weight of a passenger-pilot. That is, the Wrights figured, as long as they used efficient lifting surfaces and propellers. But of course no one had ever produced such a propeller. Marine propellers were useless and little useful information could be derived from their design. So the Wright brothers used their wind tunnel again to generate the data they needed to design the first effective airplane propeller, which they envisioned, not as a screw, but as a rotary wing. It was one of their most original scientific achievements.

They returned to Kill Devil Hills in 1903, this time to mount the engine on their new 40-foot, 605-pound *Flyer*, sporting double tails and elevators. The twin pusher propellers were mounted at the back of the plane, spinning in opposite directions for balance and driven by the engine with chains. The windswept sand dunes near Kitty Hawk were especially ideal since the Wrights constructed their biplanes with skids, not wheels. On takeoff, the plane's skids were carried on a trolley that ran along a prepared wooden track or launching rail. Handling trucks supported the wings on either side when the craft was not on the runway.

A *pylon on Big Kill Devil Hill celebrates the achievement of Wilbur and Orville Wright and marks the spont from which many glider trips began before the Wrights were ready for an engine and powered flight.* (Courtesy National Park Service)

Finally, on December 14, everything was ready. Wearing suits and ties in honor of the occasion, the two brothers tossed a coin to see who would fly the first powered aircraft. Just nine days before, another aviation hopeful, American Samuel Langley, had tried and failed in front of a flock of newspaper reporters. The *New York Times* had scoffed at him mercilessly, decried the misuse of government funding he had enjoyed and announced that success at flying was at least a thousand years in the future. Now it was the Wright brothers' turn. Wilbur won the toss for the first flight, but he lost out in the history books. As the plane took off from the wooden launching rail, he oversteered with the elevator. The plane rose too steeply, the engine stalled and *Flyer* hurled to the ground. Fortunately, the damage wasn't too bad and the Wrights were already experienced in quick repairs. Three days later they were ready to try again.

The morning of December 17, 1903 was bitter cold and overcast at Kitty Hawk, North Carolina. But the dream of centuries was about to come true. On that day at 10:30 in the morning, Orville Wright climbed aboard the flimsy little aircraft and took over the controls. Powered by its gasoline engine, the rickety structure of wood, cloth, and chains sped down the launching rail for nearly 40 feet and then took to the air. Wilbur ran alongside, supporting the wing until the plane took off. Orville headed *Flyer* into a blustery, cold 27-mph head wind. It didn't fly high: It was a scant 10 feet off the ground. And it didn't fly far: That first flight was only 120 feet in distance and lasted only 12 seconds. The flight was unruly, and Orville, too, had trouble with the controls. The aircraft's air speed of 34 mph, offset by the head wind, left *Flyer* with a ground speed of only 6.8 mph. But at last a human truly flew a heavier-than-air, powered, and controlled aircraft. The age of flight had begun.

The second flight was made by Wilbur, 175 feet in 12 seconds. The third was Orville's again, a flight of 200 feet in 15 seconds. With each flight the pilots learned more about using the controls. The fourth and final flight of the day, flown by Wilbur, lasted 59 seconds and covered 852 feet. (The landing sites for all four flights are marked on the grounds of the memorial.) While the brothers were preparing for yet another trial, though, a sudden wind squall flipped the plane over as it sat on the ground, causing damage that brought the day's work to an end. Little attention was paid to the day's events—only a handful of newspapers carried the story after so many unsupported claims by other would-be flyers. Only one photo was taken as the plane left the ground. But a barrier had been broken that morning, and the world was changed forever.

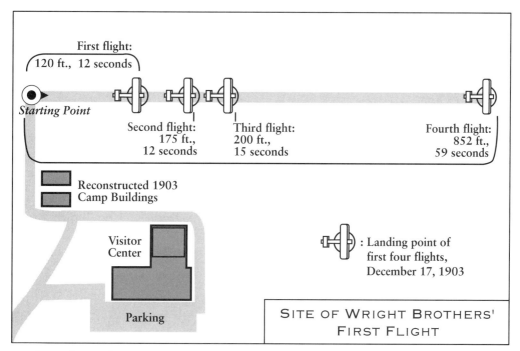

Markers indicate where each one of the first four flights ended.

The Wrights calmly returned to Dayton, where they kept working on a new design, and the following year, they flew their new plane in a 90-acre pasture outside of town. *Flyer No. 2*, with improved rudders and engine, could sustain flight for five minutes and flew turns and circles. Twice that year the Wrights invited reporters to come see what they were doing, but, strangely, the engines failed to start both times. Some biographers believe that the Wrights, who preferred to experiment in private, intentionally rigged the failures to discourage public interest—at least until they were ready.

But in 1908 both brothers proved what they had done, Orville in America, Wilbur in France. In a demonstration in Le Mans, France, Wilbur flew a circle maintaining good control and executed a gentle landing, proving that the Wrights had mastered flight. While French aviators had made considerable progress, including one 20-minute flight, they still had problems with stability and control. As one French pilot put it, "We are as children compared with the Wrights." By the end of that year the Wright brothers had logged 36 hours 20 minutes of flight time—about six times the records of all other aviators combined. Also in 1908 the U.S. government contracted with the Wrights to build an aircraft that could fly 40 mph and travel 125 miles with a pilot and passenger aboard. In the following year the two brothers incorporated the American Wright Company.

The Wrights' planes, unfortunately, included a few design elements that blocked the brothers from developing much further—since they were reluctant to change a design that worked. Four years later, on May 30, 1912, Wilbur Wright died of typhoid.

By 1914 his brother Orville left the business. Although he consulted for the Aviation Service of the Army Signal Corps during World War I, he concentrated primarily on private research. He died January 30, 1948.

The same year, on the 45th anniversary of the first controlled, powered flight ever made by humans, *Flyer No. 1* was hung in the Smithsonian Institution. And the names of Wilbur and Orville Wright, who designed, built, and flew it, will always remain the first and greatest names in the history of flight.

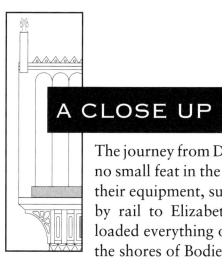

A CLOSE UP | MAKING CAMP ON THE DUNES

The journey from Dayton to North Carolina's Outer Banks was no small feat in the years 1900 to 1903. The Wrights boxed up their equipment, supplies, tools, and aircraft and shipped it all by rail to Elizabeth City on Albemarle Sound. There they loaded everything on a sailboat and sailed across the sound to the shores of Bodie Island.

The tests would take several weeks, and they would need a place to keep their aircraft and a shelter for themselves. With the help of two local men, they built temporary camp buildings of rough-hewn pine on the sand dunes

The 1903 machine and camp buildings. Wilbur Wright is standing just inside the hangar building, November 24, 1903. (Courtesy National Park Service)

near Kill Devil Hills. They left the buildings standing when they left each year, but the winter's storms took their toll, and some repairs and rebuilding were always necessary. Today visitors can see the reconstructed camp buildings, as they were in 1903, at the culmination of the brothers' experiments.

Wilbur and Orville built two buildings, side by side: a hangar to house their aircraft and protect it from the wind, and their living quarters, which they jokingly called their "Summer Home." This not-so-swank resort, they liked to claim, had five rooms: a living room, dining room, bedroom, kitchen, and workshop. In fact it was one large room, with a dresser and mirror, a table and chairs, shelves for canned goods (since they had no refrigerator, of course, and no supermarket nearby), a gasoline stove for cooking, a wood-burning stove for warmth, and four or five kerosene lamps for light in the evenings. The two men nailed burlap bags across the rafters overhead to form sling-like bunks, which they reached using a ladder. Reasoning that heat rises, they figured that this would be the warmest place in their shack for sleeping. At the back of the living quarters they had a workshop area, the entire back wall of which opened up so they could bring pieces of equipment inside to work on. They lined the walls and adjoining workbench with tools of their trade—mostly wood-working tools, such as hammers, chisels, saws, and drills. They also had tin snips, wire cutters, extra chains, a solder gun, and an acetylene torch.

Next door, the hangar measured about 44 feet long, 16 feet wide and 8 or 9 feet tall. At either end they could open the walls outward, swinging them up and propping them with poles so they could move the aircraft in and out. A narrow workbench and nails for tools ran the length of the stationary walls.

It's easy to imagine Wilbur and Orville Wright in these two buildings at the dawn of the 20th century, laboring into the night on their gliders and aircraft, rigging a new camber or repairing damage from a botched trial. Or resting in their "Summer Home," recuperating from the day's long, often disappointing experiments, figuring new answers to the problems they faced and storing up fresh hopes for the flights of the days ahead.

PRESERVING IT FOR THE FUTURE

The Visitor Center houses exhibits that are available year-round and include reproductions of the Wright brothers' wind tunnel, the 1902 glider, and the 1903 *Flyer*. Reconstructions of the original Wright camp buildings are a two-minute stroll from the Visitor Center. Adjacent to the camp buildings are granite markers that designate the lengths of the four successful powered flights. A 60-foot granite monument, dedicated in 1932, is perched atop 90-foot-tall Kill Devil Hill commemorating the achievement of these two visionaries from Dayton, Ohio.

Visitors can explore the reconstructed camp buildings and first flight trail area, and climb one of several paved paths up Kill Devil Hill to view the memorial pylon. In addition there are museum exhibits and ranger-conducted programs. (What was once a shifting sand dune is now stabilized with grass.)

Additionally, the 3,000-foot First Flight Airstrip was added to the park in 1963 to enable pilots and their passengers to visit the site of the world's first successful powered flights. (Call ahead regarding restrictions.)

EXPLORING ◆ FURTHER

Books about the Wright Brothers

Aaseng, Nathan. "The Wright Brothers and the Airplane," *Twentieth-Century Inventors*. New York: Facts On File, Inc., 1991.

Bilstein, Roger E. *Flight in America: From the Wrights to the Astronauts*. Baltimore: The Johns Hopkins University Press, 1987.

Franchere, Ruth. *The Wright Brothers*. New York: Harper & Row Junior Books Group, 1972.

Sabin, Louis. *Wilbur & Orville Wright: The Flight to Adventure*. Mahwah, N.J.: Troll Associates, 1983.

Wright, Orville. *How We Invented the Airplane*. New York: McKay, 1953.

Related Places

Greenfield Village
P.O. Box 1970
20900 Oakwood Blvd.
Dearborn, MI 48121-1970
(800) 835-5237 or (313) 271-1620
http://www.hfmgv.org

The Ohio home of Wilbur and Orville Wright, along with their cycle shop, was moved from Dayton, Ohio to Greenfield Village, adjacent to the Henry Ford Museum, in 1938.

Dayton Aviation Heritage National Historic Park
22 South Williams Street
P.O. Box 9280 Wright Brothers Station
Dayton, Ohio 45409
(513) 225-7705 (Wright Cycle Co. Building)
(513) 224-7061 (Dunbar House State Memorial)
(513) 293-3688 (Carillon Historical Park)
(513) 257-5535, Ext. 254 (Huffman Prairie Flying Field)

This park's four units preserve buildings and objects associated with the development of human-controlled, heavier-than-air, powered flight by Orville and Wilbur Wright, as well as the home of their friend, African-American poet and author Paul Laurence Dunbar. *Wright Flyer* III, flown by the Wrights at Huffman Prairie in 1905, is housed at Carillon Historical Park, and visitors can also see the place where they flew and established their flying school at Huffman Prairie Flying Field on the Wright-Patterson Air Force Base.

Rachel Carson Homestead

BIRTHPLACE OF ENVIRONMENTALIST RACHEL CARSON
Springdale, Pennsylvania

AT A GLANCE

Built: ca. 1845

Home of Rachel Carson 1907–1929

A five-room wood-frame farmhouse, birthplace and childhood home of
Rachel Carson, biologist and environmentalist.

Address:
Rachel Carson Homestead
613 Marion Avenue
Springdale, PA 15144
(412) 274-5459
http://www.rachelcarson.org

> Ecologist Rachel Carson, author of the controversial 1962 best-seller Silent Spring,
> helped launch the modern environmental movement. She began her days in this
> modest frame house, surrounded by woods, fields, streams, and creatures that
> she grew to love—and for whose continued existence she fought.

[Scientists] are never bored. We can't be.
There is always something new.
—Rachel Carson

♦ ♦ ♦ ♦ ♦

With a rare combination of scientific accuracy, artistic creativity, and passion, Rachel Carson (1907–64) awakened millions of people through her writing to the beauties of nature and its creatures. She took her readers on vicarious trips into the lives of moon jellies and starfish

Entrance to the Rachel Carson Homestead   (Courtesy Rachel Carson Homestead)

in tidepools and showed them the wonders of their biological processes and forms. She made them aware of details in the natural environment—moss and pine needles underfoot, melodic bird calls, sunshine filtering through the trees, the crispness of the night air. And as she saw that humankind was threatening the fragile balance of nature, she spoke out powerfully in a clarion call for a new sense of responsibility for our environment.

When she began life May 27, 1907, Rachel Carson's brother Robert was eight years old, and her sister Marion was ten. With such a large gap between her age and theirs, Rachel grew up spending much of her time either with her mother, who often read aloud to her, or playing alone among the wooded hills and open fields around the house where her family lived in southwestern Pennsylvania. Carson later said that she could recall "no time when I wasn't interested in the out-of-doors and the whole world of nature. Those interests, I know, I inherited from my mother and have always shared with her."

Although her family lived just a few miles east of Pittsburgh—one of the most industrialized cities in the United States—the world Rachel grew up in was rural. In 1900, her father, Robert Warden Carson, an insurance agent and land developer, had bought the 65-acre farm nestled in a bend of the lower Allegheny River among the foothills of the Allegheny Mountains, and he and her mother, Maria McClean Carson, had moved into its unpretentious little 55-year-old white clapboard house. They never farmed the land—they hoped to subdivide and sell it at a profit—but they kept a cow, chickens, rabbits, pigs, a few pets, an apple orchard, and a grape arbor. Here, in this rustic setting, is where Rachel spent her childhood and youth. She came to love this verdant setting filled with life, as well as the oceans, rivers, and streams she encountered and explored later in life, and she spent her life passing on to others the wonder and respect she had learned for the environment and its delicately balanced systems.

Money was scarce for the Carsons, the house was small, and amenities few. Their home had no indoor plumbing, and an outhouse was located in the backyard. Water for the house came from a well inside the springhouse (which was used for cold storage). This small white building still stands in the front yard, and behind the springhouse you can still see the slab for the pump. Members of the family carried water in from the well for all the cooking and washing. The original farmhouse, heated by fireplaces, had two bedrooms upstairs and two rooms downstairs, separated by the narrow stair-

The house when Carson lived there was about half this size. Restoration to its original appearance is planned for the future. (Courtesy Rachel Carson Homestead)

case—a crowded living space for five people. Visitors especially notice the low ceilings and the tiny size of the rooms. A rustic cellar beneath the house was probably used for storing canned fruits and vegetables.

Rachel's mother earned extra money by giving piano lessons on the serviceable Pickering upright where a similar one now stands against one wall in the living room. The other downstairs room was used as a dining room, and a small lean-to kitchen was added on at the back of the house. Not everything appears today exactly as it did when Rachel was a child—the house is still in the process of restoration, and further additions made in later years, after the Carsons no longer lived there, make the house appear bigger today than it was then. But future restorations will return the house to its original appearance, and in the meantime visitors exploring the original portions of the house get a feel for how cramped the house must have been, especially when Rachel's sister or brother returned to live there with spouses and children, as happened on more than one occasion. Rachel's mother encouraged her interest in the outdoors, and it's easy to see why she would have preferred it to staying indoors.

Carson loved reading and writing. By the time she was 10 she had her first article published by *St. Nicholas*, a popular magazine for children, winning its Silver Badge, a $10 prize. She decided then that she would become a writer, and much of her writing even then was about her love of nature. An article she published in *St. Nicholas* at age 14, "My Favorite Recreation," recounts a walk in the woods. With lunch box, camera, and canteen in hand and her dog Pal at her side, she takes the reader on a delightful journey observing the nests of orioles, bobwhites, cuckoos, and hummingbirds, and brings to life their trilled melodies amidst the golden rays of the sun.

After graduating from high school, Carson earned a $100 scholarship to the Pennsylvania College for Women (now Chatham College) in Pittsburgh. Although pursuing higher education was an unusual decision for a woman at the time, her parents supported it and tried to find the money to finance it. Her father sought to subdivide the land he owned, as he had planned, and, although that venture didn't go through, he did succeed in selling some of the land to pay for Rachel's education.

Intent upon a career as a writer, she began with a major in English, but reconsidered when a required biology course sharpened her already great interest in the natural world, with field trips in the woods near Pittsburgh, and explorations of nearby forests, streams, and wildlife. Awakened in her dormitory one night by a storm, she recalled a line from Alfred Tennyson's poem, "Locksley Hall": "For the mighty wind arises, roaring seaward, and I go." From that moment she felt compelled to focus on zoology instead of writing (not realizing the two could be combined). At the same time, she also felt drawn to the sea, which had always fascinated her even though she had never seen the ocean. Although she was already in her junior year, she happily threw herself into long hours of laboratory work to make up for lost time and graduated magna cum laude in 1928.

That summer, she spent six weeks of study at the prestigious Marine Biological Laboratory at Woods Hole, Massachusetts, where she soaked up knowledge from sunup to late at night, surrounded by some of the best marine biologists and oceanographers in the country. She worked in state-of-the-art laboratories and delved into the extensive libraries. And she spent hours of time at the ocean, where for the first time, she felt the rush of salt water and tiny sea creatures swirl about her. As she would later write, she "stood knee-deep in that racing water and at the time could barely see those darting silver bits of life for my tears." That fall she began work on her master's degree in marine zoology at Johns Hopkins University in Baltimore, moving her family from the cramped home she grew up in to live with her in a house she found near the Chesapeake Bay. Carson worked her way through school by teaching or as a graduate assistant, receiving her degree in 1932. After that she continued teaching to try to help her family scrape together enough money during the difficult years of the Great Depression.

The economic picture became even more bleak in 1935, when her father died of a heart attack, and Carson became her mother's sole means of support. Recalling a conversation with Elmer Higgins at the United States Bureau of Fisheries in Washington, she decided she'd call on him, even though he'd warned her that the bureau had never hired a woman scientist. Her timing, however, was excellent. The bureau was in the midst of producing a series of radio shows on fishery and marine life, which they called "Romance Under the Seas," but it wasn't going well. Could Carson write? Higgins wanted to know. She said yes; he hired her part time; and so began a 16-year career working for the Bureau of Fisheries.

The following year, after scoring highest in competition with an exclusively male group of applicants in a civil service examination, she obtained a promotion—an appointment as junior aquatic biologist, becoming the first female biologist ever hired by the Bureau of Fisheries.

Her first literary breakthrough came, strangely enough, from a government assignment, thanks to a suggestion made by Higgins. An introduction she wrote for a bureau brochure was too literary, he thought, for government documentation, but he suggested she submit it to the *Atlantic Monthly*. She did, and it was published as "Undersea" in the September issue, whetting her appetite once again for the writing life—which she began to see as a way to share a subject she loved.

The publication also snowballed into a book contract with Simon and Schuster, and Carson's first book, *Under the Sea-Wind* came out in 1941. Both literary and scientific critics gave the book high marks, but the nation's attention was turned to other matters as the bombing of Pearl Harbor plunged the United States into war, resulting in poor book sales. After the end of World War II, thoughtful people everywhere, including Rachel Carson, began to consider the far-reaching implications of nuclear warfare. Highly destructive nuclear bombs had been dropped on the cities of Hiroshima and Nagasaki in Japan, leaving behind not only a path of destruction but radiation that continued to harm living things for years to come. Carson had always thought, like many others, "that the stream of life would flow on through time in whatever course God had appointed for it . . . without interference from one of the drops of the stream—man." Now she began to recognize she had been wrong. It was a thought that would grow with her, culminating later in a book that would change the way the world thought about humanity's effect on the environment.

By 1947 Carson began to think about writing another book, not just for scientists, but "for anyone who has looked out upon the ocean with wonder." By now she had risen to the position of assistant to the chief of the Office of Information and soon afterward, in this exclusively male domain, achieved the grade of biologist and editor in chief of the Information Division for the newly created Fish and Wildlife Service.

She had the opportunity to go on an underwater diving expedition, becoming the first woman to sail the northern Atlantic aboard a Fish and Wildlife Service research boat, and her strenuous 10-day research trip gave her a new sense of the vast size and endless movement of the ocean. As she was lowered down through ocean layers in the bulky diving suit, she saw astounding views through the windows of her diving helmet. And she was struck anew by the wonder of the sea full of animals living amidst strong ocean currents—an ocean even more powerful and more alive than she had imagined it.

All of this went into her new book, *The Sea Around Us* (1951), which she described as a "biography of the sea." A pre-publication chapter appeared in the *Yale Review* as "The Birth of an Island," for which she won the George Westinghouse Science Writing Award, and shortly thereafter parts of the book were reprinted by the *New Yorker, Nature Magazine* and *Reader's Digest*.

Her agent saw that re-release of *Under the Sea-Wind* might bring additional sales, and both books wound up on the best-seller list at the same time, a phenomenon described by the *New York Times* as "rare as a total solar eclipse." *The Sea Around Us* remained there for 81 weeks—more than a year and a half. It was translated into 32 languages, won her a National Book Award, and prompted her election to the British Royal Society of Literature.

The success gave Carson the financial independence she needed to write full time. She left her position at the Fish and Wildlife Service and built a house on the coast of Maine, where she and her mother moved. Immediately Carson set to work on a new book, a detailed look at organisms living along the shoreline near her new home and down the coast. In it she explored her philosophy of ecology, as well as her love of nature. "The shore is an ancient world," she said, "Each time I enter it, I gain some new awareness of its beauty and its deeper meanings, sensing the intricate fabric of life by which one creature is linked with another, and each with its surroundings." She called the book *The Edge of the Sea* (1955), and this work, too, reached the best-seller list, enjoyed *New Yorker* excerpts and a *Reader's Digest* condensed version.

But the work for which Carson wanted most to be remembered was *The Sense of Wonder*, which encouraged adults to share their love and appreciation for the natural world with children. Originally called "Help Your Child to Wonder," a 1956 article written for *Woman's Home Companion*, this work promoted a cause Carson embraced with special enthusiasm. Shortly after her father's death, her sister Marion had died at the age of forty, leaving two young daughters, and Rachel and her mother had taken the two girls in. So, even though Carson never married, children had played an important part in her life. In fact, only two years after publication of the article, her niece died and Carson legally adopted her grand-nephew, Roger, who was five years old. Carson's mother died the following year at the age of 88. Carson had always planned to expand the article into a book but never had time to do that. *The Sense of Wonder* was published posthumously, in 1965, and its theme plays an important part in the Rachel Carson Homestead's interpretive mission.

But the work for which Rachel Carson is best known might never have been done had it not been for a letter she received from a friend, Olga Huckins, who lived in Duxbury, Massachusetts. A plane had passed overhead spraying DDT one morning in the summer of 1957. The following day Huckins found seven dead songbirds in her yard; she made a connection between the oily spray from the plane and the dead creatures and she wrote an angry letter to the newspaper, sending a copy to her friend. The story raised a flag in Carson's mind, reminding her of research done years earlier by Elmer Higgins and another colleague at the Fish and Wildlife Service on the effects of chlorinated hydrocarbons, such as DDT, on wildlife. She had tried then to interest *Reader's Digest* in the story, but had been turned down, and the results were never published.

Carson began looking into the question again. She got in touch with other biologists, chemists, and geneticists, and she got back a mound of documentation. She investigated legal suits—complaints lodged by farm workers who had become sick and by owners of pets and livestock that had suffered poisoning. The more she looked, the more frightening she saw the evidence was. The result of her research was not an article; it was a book.

DDT had first come into use in 1939, when Swiss chemist Paul Müller found that it was extremely toxic to insects while apparently harmless to humans. Its use became extremely effective against devastating tropical diseases carried by insects, such as malaria and typhus. And it became widely used by farmers as the "miracle tool" for destroying harmful insects. In 1948, Müller received the Nobel Prize for his work.

But Carson began to put together another picture. Insects had begun to develop resistance to DDT, so that higher and higher doses were necessary to produce a lethal effect. By 1949, resistant houseflies and mosquitoes had both appeared. Larger animals, such as the birds found by Olga Huckins, were also affected. The smaller ones were killed outright; larger ones retained the poison in their fatty tissues and passed it on in milk and meat. Eggshells became thinned to the point that young birds were crushed before they even hatched. Birdwatchers and fishermen had begun to notice that bird and fish populations were growing smaller, and DDT seemed a likely culprit. Each spring, the sounds of nature were becoming less and less prevalent. Carson began to draw a picture of sickness and death, describing a "strange stillness" replacing the songs of birds.

Carson found that toxins sprayed to eradicate sage in favor of grasslands for grazing also destroyed willows that provided habitats for moose, beaver, and other animals. Without the willows, the land became parched and dry. Without the beavers, the lakes created by their dams ran off and the trout population diminished. Were the gains in grassland worth the risk to the rest of the ecological balance?

The topic was controversial, and Carson knew it. Ecology was not a commonly held concept at the time, and neither was environmental protection. Always meticulous about her facts, she checked and rechecked her data and conclusions. She honed her writing, wrote and rewrote. The project took her four years to complete, during which time she was besieged by illnesses, including angina, arthritis, and breast cancer. And when the book was finally finished, only the *New Yorker* had the courage to pre-publish excerpts of the book she called *Silent Spring*. But whe she turned in her manuscript, editor William Shawn called personally to tell her, as she recounted in a letter to her friend Dorothy Freeman, that it was "a brilliant achievement." "You have made it literature," he told her, a piece "full of beauty and loveliness and depth of feeling." She knew then that she had done what she had set out to do.

But the work was also a bombshell. The press called Carson "hysterical" and her work "unscientific." Chemical firms, farmers, ranchers, and housewives all had found pesticides a boon. So had public health officials and disease control experts. Her news was very unwelcome, and some well-reasoned arguments still maintain that she overstated the case. But a handful of influential people heard her message and supported her, including scientist/writers Loren Eiseley and Julian Huxley. President John F. Kennedy consulted his Science Advisory Committee, which published a report concerning pesticide use and control, made public May 15, 1963, that confirmed every point Carson had highlighted in *Silent Spring*. The following day a congressional subcommittee began an investigation of government and industry regulations regarding pesticides.

Rachel Carson had once said, "I could never again listen happily to a thrush song if I had not done all I could." Now she knew she had. She died April 14, 1964 of breast cancer at the age of 56.

Seven years after the publication of *Silent Spring, Time* magazine, which had initially berated the book, began running a weekly "Environment" section, running Carson's picture in the first issue. In 1970, 4,000 acres of southeast coastal marsh in Maine was dedicated as the Rachel Carson National Wildlife Refuge. The United States Postal Service issued a commemorative stamp in her honor in 1981 as part of its Great Americans series. And today, most textbooks on science mention her contribution to environmental science.

In her writings about the sea, Carson showed her readers how all life was interdependent—inextricably tied together. In *Silent Spring*, she showed what enormous influence we have on everything on our planet, how the very future of all life depended on our wise decisions. Her love of nature and the beauty she captured in her writing helped humans the world over look at life on their planet in a new and more responsible way.

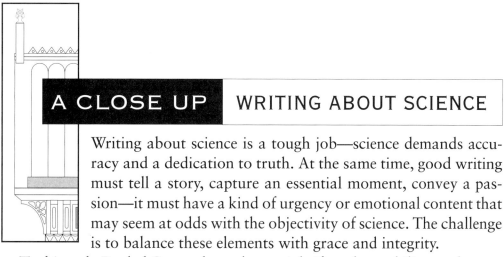

A CLOSE UP   WRITING ABOUT SCIENCE

Writing about science is a tough job—science demands accuracy and a dedication to truth. At the same time, good writing must tell a story, capture an essential moment, convey a passion—it must have a kind of urgency or emotional content that may seem at odds with the objectivity of science. The challenge is to balance these elements with grace and integrity.

To this task, Rachel Carson brought special gifts: a keen ability to observe, an intense love of knowledge, and a fine talent for description. To these she added an inherent honesty that caused her coworkers at the Fish and Wildlife Service to describe her as "like Gibraltar." Wildlife artist Bob Hines, a close friend and associate who worked with her there, once said, "She had no patience with dishonesty or shirking . . . she didn't like shoddy behavior . . . she had standards, high ones." And she applied those standards, above all, to herself and her work. Carson also had an abundant zest and curiosity, as well as the great patience to probe until she found the answers to questions, even if the answers weren't popular.

But Rachel Carson's unique mark as a writer was her romantic and passionate view of her subject matter—of which she made a gift to her readers, as she did in this description of a Georgia beach at twilight from her book *The Edge of the Sea*:

Once, exploring the night beach, I surprised a small ghost crab in the searching beam of my torch [flashlight]. He was lying in a pit he had dug just above the surf, as though watching the sea and waiting. The blackness of the night possessed water, air and beach. It was the darkness of an older world, before Man. There was no sound but the all-enveloping primeval sounds of wind blowing over water and sand, and of waves crashing on the beach. There was no other visible life. Just one small crab near the sea.

PRESERVING IT FOR THE FUTURE

The Rachel Carson Homestead Association, a private nonprofit organization, was formed in 1975 to preserve and restore Rachel Carson's birthplace and to provide environmental education programs that advance her environmental ethic.

Visitors from all over the world, such as this group from Japan, come to pay homage to Rachel Carson at the homestead site. (Courtesy Rachel Carson Homestead)

Today, the homestead offers guided tours of the house and grounds, environmental education classes, school and outreach programs, and it also serves as an international resource for information about Carson's life and work. A wooded area nearby has been made available for the association's use and visitors can explore a ¼-mile nature trail there, following a self-guided interpretative hike based on Carson's childhood story, "My Favorite Recreation," written about her own walks in the woods.

In 1997–98, plans call for removal of portions of the house that were not present in Carson's time.

EXPLORING ◆ FURTHER

Works by Carson

Carson, Rachel. *The Edge of the Sea*. Boston: Houghton Mifflin, 1955.

———. *The Sea Around Us*. New York: Oxford University Press, 1951.

———. *Silent Spring*. Boston: Houghton Mifflin, 1962.

———, and Dorothy Freeman. *Always, Rachel: The Letters of Rachel Carson and Dorothy Freeman, 1952–1964*. Edited by Martha Freeman. Boston: Beacon Press, 1995.

Works about Rachel Carson

Gartner, Carol B. *Rachel Carson*. New York: Ungar Publishing Co., 1983.

Jezer, Marty. *Rachel Carson*. New York: Chelsea House, 1988.

Kudlinski, Kathleen V. *Rachel Carson, Pioneer of Ecology*. New York: Viking Kestrel, 1988.

Peterson, Robert, "Rachel Carson: Sounding the alarm on pollution," *Boys' Life* 84 (August 1, 1994), p. 38.

Sterling, Philip. *Sea and Earth: The Life of Rachel Carson*. New York: Thomas Y. Crowell, 1970.

Veglahn, Nancy. "Rachel Carson (1907–1970)," *Women Scientists*. American Profiles Series. New York: Facts On File, 1991.

Related Place

Rachel Carson National Wildlife Refuge
321 Port Road
Wells, ME 04090
(207) 646-9226

Established in Rachel Carson's honor to preserve wildlife habitat and critical waterfowl migration routes, the refuge's ten divisions stretch across approximately 7,000 acres in southern Maine—between Kittery and Cape Elizabeth. Of this area, a one-mile interpretive trail, the Carson Trail, is open to the public, seven days a week, from dawn to dusk. Located at the headquarters in Wells, the trail follows the edge between a mature upland woodland and a tidal marsh. Visitors can pick up a self-guiding interpretive leaflet at the headquarters (Monday through Friday 8:00 A.M. to 4:00 P.M.) to aid in understanding this beautiful environment and its habitats, where woodland songbirds and wading birds can be seen, as well as occasional waterfowl, salt marshes, and grasses.

U.S. *Space and Rocket Center*

BIRTHPLACE OF ROCKETS TO THE MOON
Huntsville, Alabama

AT A GLANCE

Built: 1970

The U.S. Space and Rocket Center, adjacent to Redstone Arsenal,
the U.S. Army's missile and rocket center and the National Aeronautics and
Space Administration's (NASA's) George C. Marshall Space Flight Center.

At the Redstone Arsenal, between 1950 and 1960, Wernher von Braun built
the rocket that launched America's first satellite. At the Marshall Space Flight
Center, established in 1960, engineers developed the *Saturn* V rocket that
carried astronauts to the Moon. These historic events are captured at the
U.S. Space and Rocket Center at the same location.

Address:
NASA Marshall Space Flight Center
Public Inquiries/CA20
Huntsville, AL 35812

Alabama Space and Rocket Center
Huntsville, AL 35812
(800) 633-7280 or (205) 837-3400

An *extraordinary team of scientists, led by rocket scientist Wernher von Braun,
built America's first space rockets at this site. Visitors can see replicas of the rockets
they built, actual spacecraft, test stands, and the inner workings of
current NASA projects.*

Sometimes you can take one look and see something
obviously wrong—not accessible perhaps, or too flimsy.
The people who are working with it all day are too close to
see it. That's why I go to the fabricating shop—I want to
know what my baby will look like.

—Wernher von Braun,
about his work at
Marshall Space Flight Center

♦ ♦ ♦ ♦ ♦

People from Huntsville, Alabama like to call their hometown "Rocket
City," and with good reason. It is the home of one of the most
monumental pieces of technology ever built by humankind, the *Saturn
V* rocket. This huge construction of steel, with its massive engines and
powerful thrust, provided the propulsion that made possible the Apollo
flights to the Moon and back. Anyone who has seen the movies *Apollo 13*
or *The Right Stuff* knows the excitement mixed with terror that accompanied
each thunderous lift-off of these giants. And this is the place where their story
began.

A visit to the Marshall Center—one of NASA's largest—takes visitors back
to the early stages of humankind's breathtaking venture into space to view
a full-size model of a *Saturn V* rocket. This was the dynamo that lifted the
45-ton Apollo spacecraft into Earth orbit. Only a rocket this mammoth could
muster the power to launch the lunar missions on their way—because
additional fuel and power were needed to escape Earth orbit and pull away
from the home planet's gravity to set off for the Moon.

Visitors also see the *Apollo 16* capsule that carried astronauts John Young, Charles Duke, and Thomas Mattingly to the Moon in 1972; a Moon rock that once lay on a surface more than 200,000 miles away; and a Lunar Rover (or Moon buggy). A full-size mock-up of the space shuttle is also on display.

The first successful American rockets were shot off in a field near Auburn, Massachusetts by a quiet physicist named Robert Goddard in the 1920s. But the biggest and most rapid advances in early rocketry began in Germany, the home of well-known Wernher von Braun, who is commonly thought of as the father of the American space program. In Huntsville he is a hero.

When Wernher von Braun was a boy in Germany, he became intrigued by the idea that people could someday escape the cradling atmosphere of the Earth and travel into the far reaches of space. Writers had long told tall tales about traveling to the Moon and other planets, but these were more fantasy than reality. No one at that time had ever figured out how to build a vehicle that could overcome the grasp of gravity that holds us earthbound, although a few people had begun to develop theories. Young von Braun read about

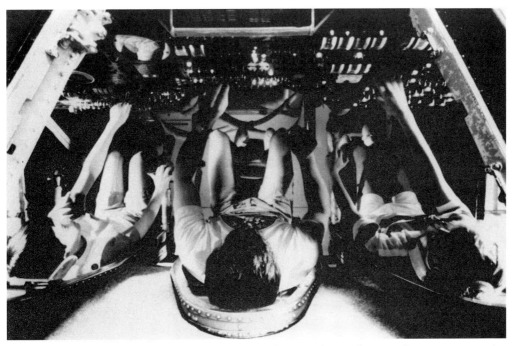

You can "pilot" your own Apollo mission inside a rocket capsule simulator at the U.S. Space and Rocket Center. (Courtesy U.S. Space and Rocket Center)

these, and by the time he was in his teens he joined an amateur rocket society that hoped to build a space rocket.

Some 25 years later, in 1950, Wernher von Braun arrived in Huntsville, Alabama with a group of other German rocket scientists to build rockets for the United States government. He had come a long way since his days as a boy when he had puzzled over the sketches of untried rocket technology that he found in magazines. What had once been science fiction was fast becoming a reality.

But the road had been rocky. Shortly after young von Braun joined the amateur rocket society in Germany, the German army under Adolf Hitler began to gear up for war. Recognizing that rocket technology had a much more down-to-earth potential in weaponry, the German army hired von Braun and many of his rocket society colleagues to develop rocket technology. Their work became designing the deadly V-2 weapon unleashed by Germany toward the end of World War II. The V-2 was a rocket capable of traveling many miles, and Germany launched some 3,225 of them, causing death and destruction throughout southern England. Its effect was so devastating that some historians believe if the V-2 had been ready for use earlier, Germany and its allies Italy and Japan might have won the war.

Instead, the V-2 rocket did not change the course of the war: In 1945 the Allied Forces—including Great Britain, France, the Soviet Union, the United States, China, and others—succeeded in defeating Germany. After that, suddenly, some of the picture changed rapidly. Recognizing a coming shift in allegiances, the United States worried about being able to protect its shores against the Soviet Union, whose communist government was basically unfriendly to democratic, capitalistic governments. The chill of the cold war, a period of threatening animosity and angry rivalry, began to set in at once.

So, as soon as Germany surrendered, the United States sought out the German rocket builders, recognizing that if they were not on the American side of the cold war, they might wind up helping the Soviets. Moving quickly, in June 1945 the U.S. government secretly flew nearly a hundred German rocket scientists across the Atlantic to test rocket technology in the New Mexico desert. By 1950, most of them, including Wernher von Braun, had become United States citizens, and the U.S. Army moved them to the new Redstone Missile Site in Huntsville. There, they would build rockets.

But in all this time, von Braun had not forgotten his childhood goal: to send a rocket into space, ultimately with a human pilot aboard. Then on October 4, 1957, Americans saw a new "star" in the sky—the Soviet satellite *Sputnik*, the first human-made satellite ever launched. In the struggle to compete for international prestige, this was a great blow to American national pride. The USSR's communist government had won the race into space. It was also a show of power: If the Soviets had rockets powerful enough to lob a satellite into orbit, certainly they had firepower enough to do considerable damage to any enemy in the world.

Von Braun quickly stepped in. He and his team already had rockets capable of putting a satellite into orbit, he told the United States government. On January 31, 1958, within a few weeks after receiving a go-ahead, he and his team successfully launched *Explorer I*. They followed that success with the launching of a satellite to explore the Sun, *Pioneer IV*, on March 2, 1959, as well as the successful flight and recovery of two animals in space, monkeys Able and Baker, on May 28, 1959. And the rockets they used were based on the same technology they had used in Germany for the V-2 rockets. Only this time they were rockets for peace.

In 1960, the newly formed National Aeronautics and Space Administration (NASA) established a new center at Huntsville, the George C. Marshall Space Flight Center, naming Wernher von Braun the director. "The Center," NASA announced, ". . . will have charge of developing and launching NASA's space vehicles and conducting related research. It is the only self-contained organization in the nation which is capable of conducting the development of a space vehicle from the conception of the idea, through production of hardware, testing, and launching operations."

As NASA embarked on a commitment to send human beings to the Moon, it became the task of the rocket team at Huntsville to build the rockets that would send them there, a job they were ready and enthusiastic about performing and performing well.

The success story of the rockets built at Marshall had its roots, von Braun would always insist, in the tireless and seamless coordination of effort by an extraordinary team, characterized, as he once wrote, "by enthusiasm, professionalism, skill, imagination, a sense of perfectionism, and dedication to rocketry and space exploration."

The center in Huntsville features the only full-scale space shuttle on display. (Courtesy U.S. Space and Rocket Center)

A firm believer in the merits of teamwork in science, von Braun told *New York Times* reporter George Barrett in an October 1957 interview: "Take a group of people with the same general fund of knowledge, present them with a problem and you will get it done. The idea of a great genius sitting quietly in some corner dreaming up vast secrets is no longer real. So many things in modern science have become so wide in scope, so intricate, that more and more it takes groups of experts to do the work."

Marshall began as a rocket-building site. Today it is that and more. This is NASA's lead center for transportation systems and research in microgravity —what happens to substances and organisms in environments having less gravity than we enjoy on Earth's surface. This is also a center for scientific research in materials, biotechnology, global hydrology and climate studies, astrophysics, and payload utilization. Today more than 3,000 people work at Marshall, with another 11,000 in private industry in the surrounding area who are counted as part of the "Marshall team."

A visit to Marshall takes you inside today's space program with a model of the Hubble Space Telescope (the space-based astronomical observatory that began in 1990 to send back images of portions of the universe never seen before with such clarity) and a tour of the areas where the International Space Station is being built, as well as the neutral buoyancy space simulator (a huge pool where astronauts can practice working in simulated weightlessness), the propulsion and structural test facility, the Redstone test stand that was used in the 1950s, the *Saturn V* dynamic test stand, and a replica of the *Saturn V* launch vehicle. Here, too, you can see a full-size model of the Redstone rocket that launched America's first satellite, as well as *Pathfinder*, NASA's full-scale mock-up of the space shuttle, which was used for testing during the shuttle development program—representing the full gamut of NASA's ventures into space, from the beginnings to the present and into the missions of tomorrow.

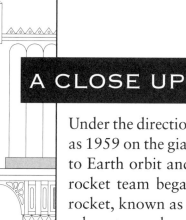

A CLOSE UP — THE SATURN V ROCKET

Under the direction of Wernher von Braun, work began as early as 1959 on the giant *Saturn V* rocket that could lift heavy loads to Earth orbit and send them on their way to the Moon. The rocket team began with development and testing of a small rocket, known as *Saturn I*, which they then modified to create a booster rocket—one that had enough power to provide a "mid-range" boost. Called Saturn IB, this rocket used much of the same technology that would go into the *Saturn V*, without the enormous size and power.

Saturn IB was used extensively for testing—launching early models of the main spacecraft, the Apollo Command Module (CM), into Earth orbit. Because the rocket and spacecraft were being developed simultaneously, the tests served a dual purpose: testing the main spacecraft the astronauts would travel in and, at the same time, testing the rocket technology. In 1966, three Saturn IB rockets launched unmanned tests, without astronauts on board, and a manned flight was planned for early 1967. But during a pre-flight test of the CM, an electric arc ignited the oxygen atmosphere inside the module, and astronauts Gus Grissom, Roger Chaffee, and Edward White met their deaths trapped in the fire inside the tiny spacecraft. Until the cause of the fire could be eliminated and stricter safety precautions set in place, all testing came to a standstill for nearly a year.

When the first *Saturn V* was launched November 9, 1967, carrying an unmanned Apollo spacecraft into orbit, it passed the test with a thunderous and reassuring roar. The rocket's three stages went off perfectly as planned. The first, composed of five engines in a cluster at the base of the huge,

Wernher von Braun using a periscope to check on a rocket launch at Kennedy Space Center
(NASA Marshall Space Flight Center Archives)

6.2-million-pound rocket, lifted the giant structure with a deafening roar that shook the ground with its might. Well up into the atmosphere, the second stage took over as the second cluster of five engines began to burn. Finally, the third stage came into play. The restartable engine of the third stage made it possible during Moon missions to park the Apollo in a temporary Earth orbit at 17,400 mph, then switch off and reignite to head toward the Moon.

From 1967 to 1972, the *Saturn V* rocket launched 12 more Apollo missions—one more unmanned test flight and 11 manned Apollo missions (two in Earth orbit, three around the Moon, and six Moon landings). In 1972 the last *Saturn V* rocket lift-off carried a 120-ton laboratory called *Skylab* into Earth orbit, where astronauts conducted a series of scientific missions.

At the U.S. Space and Rocket Center, a replica of the *Saturn V* lies on the ground, so you can walk its entire length—about as long as a football field. It's impressive to realize that not once, with all its millions of parts, did the great giant ever fail.

PRESERVING IT FOR THE FUTURE

In 1968, as director of Marshall Space Flight Center, Wernher von Braun approached the Alabama legislature with the idea of creating a museum jointly with the U.S Army Missile Command and NASA. Financing for the construction was approved by the state's lawmakers and voters in 1968, and the U.S. Army donated land on its Redstone Arsenal, where Marshall Space Flight Center is also located. The museum opened its doors in 1970.

The Space Center houses more than 1,500 pieces of rocket and space hardware, including some 300 major artifacts on loan from the Smithsonian Institution's National Air and Space Museum in Washington. An exhibit on Wernher von Braun's life explores the human side of the quest for the Moon, which, as curator James Hagler remarked, "It's a great history lesson and demonstrates how von Braun was a man ahead of his time." The space museum is also the visitor's information center for the Marshall Space Flight Center and for the U.S. Army Missile Command at Redstone Arsenal, and visitor tours of NASA begin here.

EXPLORING ◆ FURTHER

Books about Space Exploration and Rocketry

Fries, Sylvia Doughty. *NASA Engineers and the Age of Apollo*. Washington, D.C.: National Aeronautics and Space Administration, 1992.

Murray, Charles, and Catherine Bly Cox. *Apollo: The Race to the Moon*. New York: Simon and Schuster, 1989.

Ordway III, Frederick I., and Mitchell R. Sharpe. *The Rocket Team*. Foreword by Wernher von Braun. New York: Crowell, 1979.

Spangenburg, Ray, and Diane K. Moser. *Space Exploration: Opening the Space Frontier*. New York: Facts On File, Inc., 1989.

———. *Wernher von Braun: Space Visionary and Rocket Engineer*. New York: Facts On File, Inc., 1994.

Internet Connection

http://spacelink.msfc.nasa.gov/home.index.html

Related Places

Lyndon B. Johnson Space Center (for information)
2101 NASA Road One
Houston, TX 77058
(281) 483-0123 or (281) 483-8693 (for information)

Space Center Houston (to visit)
2101 NASA Road One
Houston, Texas 77058
(281) 244-2100; Fax: (281) 283-7724

Established in 1961, JSC is NASA's primary center for design, development, and testing of spacecraft for human flight, as well as the selection and training of astronauts. From Mission Control located here, NASA controlled all aspects of the Apollo Moon missions, from lift-off to splashdown. Through programs and exhibits maintained by Space Center Houston (run by a separate nonprofit organization appointed by NASA), visitors can see a full-scale mockup of the space shuttle, including flight deck and mid-deck, try on a space helmet and try out a simulation of the robotic arm (Manned Maneuvering Unit) used by astronauts on the shuttle. A tram tour behind the scenes enables visitors to see many of the historic and current facilities that support the manned space program. Although the sights vary, most tours include Mission Control, the Space Environment Simulation Laboratory, the Mock-Up and Integration Laboratory, and the Weightless Environment Training Facility (WET-F), which is similar to the one at Marshall Space Flight Center in Huntsville.

Spaceport USA
Kennedy Space Center, FL 32899
(407) 452-2121
Exploration Station: (407) 867 2959
Educator Resources Laboratory: (407) 867-4090

Located on the NASA Parkway West between the KSC Industrial Area and U.S. 1 on the mainland, this site offers a visitors center, where visitors can view films of a shuttle lift-off, planet Earth seen from space, and other topics. The Rocket Garden, an outdoor exhibit, contains authentic rockets and spacecraft, as well as a full-scale model of a space shuttle orbiter, including a walk-through exhibit. The Gallery of Space Flight exhibit hall contains a Moon rock and other historical exhibits on human space flight.

The chapters of this book explore only a few of the many historic sites that commemorate aspects of the development of science and invention. Following is a list—by no means exhaustive—of additional historical places related to science and invention in the United States. Some are fully interpreted historic parks or monuments, while others are no more than a plaque or a marker. But all recognize individuals who had the curiosity and perserverance to explore how the world works and to find ingenious ways to put what they learned to good use.

Northeastern States

Matzeliger Monument
Pine Grove Cemetery
25 Gentian Path
Lynn, MA 01905

Jan Matzeliger (1852–89) came from Dutch Guiana to the United States, where he went to work as a cobbler in the town of Lynn, Massachusetts. He invented an automatic "sole machine" that could attach a leather upper automatically to a shoe sole. It was an innovative and efficient invention, and it put many craftspeople out of work, destroying the town's industrial base. Nonetheless appreciative of his inventive expertise, the town erected a monument at the gravesite of this African-American inventor.

Lowell National Historic Park
67 Kirk Street
Lowell, MA 01852-1029
(508) 970-5000

Built on the banks of the Merrimack River north of Boston, this 19th-century mill town could well be considered the cradle of the industrial revolution in the United States. Here one of the earliest experiments with mass production took place, combining technology with a paternalistic social system for workers. Here "mill girls"—young women from neighboring farms—worked to earn extra money for their families. Closely supervised both on and off the job, they worked as many as 70 hours a week (at $2.25 to $4.00 a week) and lived in boardinghouses in town. Later, immigrants expanded the labor force, including Irish, French Canadians, Poles, Greeks, and Jews. The Lowell National Historical Park preserves the canal system built here, as well as seven of the ten original mills. Also preserved are some of the early boardinghouses, immigrant neighborhoods, and the trolley system. Guided tours are offered of the canal and mills during the summer.

Lewis Temple Statue
613 Pleasant Street
New Bedford, MA 02740

A monument commemorating the invention by African-American metalsmith Lewis Temple of a type of harpoon—known as the Temple "toggle" iron—which was used by whalers to fasten lines securely to the great whales they hunted. Although it was widely used on hundreds of whaling voyages, Temple never patented the device and died destitute. This statue was created in Temple's honor by black sculptor James Toatley.

Springfield Armory National Historic Site
One Armory Square
Springfield, MA 01105-1299
(413) 734-8551; Fax: 747-8062

Established in 1794, until 1967 this armory was the U.S. Army's main research and development center for development of small arms. Springfield Armory technology profoundly affected the lives of both soldiers and civilians. In addition to the effect of the armory's arms development on military tactics, armory inventions revolutionized the manufacture of many consumer products, including keys, shoes, baseball bats, and furniture. The site's museum is housed in the original Main Arsenal built in the 1850s. The site is maintained by the National Park Service.

Benjamin Banneker Marker
Mount Gilboa A.M.E. Church
2312 Westchester Avenue
Catonsville, MD 21228

A monument to Benjamin Banneker (1731–1806) a free black mathematician-scientist and proponent of black equality, whose almanacs based on astronomical observations were well known in his time. Banneker helped calculate the boundaries for Washington, D.C., wrote a dissertation on bees, and calculated the recurrent cycle of locust plagues.

George Eastman House
International Museum of Photography & Film
900 East Avenue
Rochester, NY 14607
(716) 271-3361

This 12.5-acre museum site was the urban estate of George Eastman (1854–1932), founder of the Eastman Kodak Company. The site includes Eastman's restored house (built in 1905) and gardens, an archives building and research center, galleries, two theaters, and an education center that offers discovery programs. The museum displays the art, technology, and impact of photography and cinema over 150 years and interprets the life of Eastman, who invented a simple camera in 1888 and marketed the first roll of film—making photography a popular pastime.

The Franklin Institute Science Museum
222 North 20th Street
(at Benjamin Franklin Parkway)
Philadelphia, PA 19103
(215) 448-1200
Internet website: http://www.fi.edu

The Franklin Institute was founded in 1824 to honor scientist/inventor Benjamin Franklin. In 1934, the institute opened its museum, which offers a wide variety of programs featuring a hands-on approach to science and technology. The museum also interprets the social and historical impact of both science and technology. Exhibits include, among others, the Benjamin Franklin National Memorial (where visitors can see the glass static tube and electrostatic machine that he used to perform his scientific experiments), Railroad Hall, and Shipbuilding on the Delaware, as well as exciting films and interactive presentations. The museum's Tuttleman Omniverse Theater shows stunning Omnimax films on topics such as "The Rainforest," "Sharks," and "Antarctica." Projected onto a four-story, domed screen with fifty-six speakers, these films give the experience of being a part of the action. Also here is the Fels Planetarium, the second planetarium constructed in the United States. The online Internet site is a rich resource for planning a visit and also provides online exhibits you can visit from home.

Steamtown National Historic Site

150 South Washington Avenue
Scranton, PA 18503
(717) 340-5200; Fax: 340-5235; TDD: 340-5207

Created by Congress in 1986, Steamtown National Historic Site interprets the story of Main Line steam railroading, from 1850 to 1950. Located in downtown Scranton, the site includes about 40 acres of the Delaware, Lackawanna, and Western Railroad's former Scranton Yards. Some of the historic railroading buildings date back to 1865, and Steamtown owns more than 30 steam locomotives and 78 cars, including cabooses, coal hoppers, baggage cars, and passenger coaches. The site operates three historic steam locomotives (including brief train rides in the yard as well as scheduled longer rides) and offers switching demonstrations in the yard. The core complex area, opened in 1995, includes the restored 1937 roundhouse, the 1902 roundhouse, a visitor center, theater, history museum, and technology museum.

Southeastern States

Benjamin Banneker Boundary Stone

18th and Van Buren Streets
Arlington, VA 22205

One of the first recognized African-American scientists, Benjamin Banneker, a mathematician and astronomer, performed the astronomical calculations necessary to position the cornerstones of Washington, D.C. when the city was first laid out. This stone is one of them and was declared a National Historic Landmark in 1976. It defined the southwestern border of the District of Columbia at the time, although it is located on a portion of land that was restored to Virginia in 1847.

Cyrus McCormick Farm and Workshop
Shenandoah Valley Agricultural Research
  and Extension Center
McCormick Farm Circle
Steele's Tavern, VA 24476
(540) 377-2255

In this workshop, McCormick invented the mechanical reaper, a machine that served to revolutionize the practice of agriculture and formed the basis for a company that would later become International Harvester. Dating from 1822, both the house and farm are preserved; a small museum recalls McCormick's achievement.

Garrett A. Morgan Historic Marker
Winchester Road and Vine Street
Paris, KY 40361

Born in the small town of Paris, 17 miles outside of Lexington, African-American inventor Garrett A. Morgan later moved to Cleveland. He invented the gas mask, the red-yellow-green traffic light, and a hair-straightening solution. The town of Paris has erected a marker in his honor.

Midwestern States

African-American Scientists from A to Z Exhibit
Children's Museum
3000 North Meridian Street
Indianapolis, IN 46208
(317) 924-5431

This permanent exhibit explores the contributions of such African-American inventors and scientists as astronomer/surveyor Benjamin Banneker, physician Charles Drew (who developed the first U.S. and British blood banks), inventor Elijah McCoy (from whose products' quality the term the "real McCoy" derives), chemist James Harris (co-discoverer of elements 104 and 105), and many others.

The Ford Museum
20900 Oakwood Blvd.
Dearborn, MI 48121-1970
(800) 835-5237 or (313) 271-1620
http://www.hfmgv.org

In addition to the Edison and Wright exhibits, this museum includes exhibits of Henry Ford's automobiles and the history of the development of their technology, along with a replica of the Detroit workshop where he built his first car in 1896.

Western States

Lake Roosevelt National Recreation Area
1008 Crest Drive
Coulee Dam, WA 99116
(509) 633-9441

The Grand Coulee Dam, located on the Columbia River in central Washington, is the largest concrete structure in the United States. One of the civil engineering wonders of the world, it produces up to 6.5 million kilowatts of power and irrigates more than half a million acres of Columbia River basin farmland. Built from 1933 to 1942, it is 550 feet high, 450 feet thick, and nearly a mile across. Visitors can see exhibits on the construction and impact of the dam in the Visitor Arrival Center and take an elevator down into the powerhouses to view the massive turbines that transform water power into electrical power.

Experimental Breeder Reactor I
National Historic Landmark
Idaho National Engineering Laboratory (I.N.E.L.)
Arco, ID 83213

In 1951 the I.N.E.L. was the site of one of the most significant scientific accomplishments of the century—the first use of nuclear fission to produce electricity. This took place at Experimental Breeder Reactor I, now a National Historic Landmark open to the public.

Jet Propulsion Laboratory
4800 Oak Grove Drive
Pasadena, CA 91109
(818) 354-4321

Founded in 1963, the Jet Propulsion Laboratory (JPL) is the heart of NASA's planetary exploration program. Managed for NASA by the California Institute of Technology, JPL has developed and directed planetary missions to all known planets in the Solar System except Pluto (and one to Pluto is planned). Visitors can tour the Spacecraft Museum, the Space Flight Operations Facility (mission control area), and the Spacecraft Assembly Facility. Tours for individuals and families by arrangement (four to six weeks in advance), and for groups, six months in advance.

Southwestern States

White Sands National Monument
P.O. Box 1086
Holloman Air Force Base, NM 88330
(505) 479-6134

This site includes the White Sands Missile Range Trinity Site, where the first atomic bomb was exploded. This test, conducted on July 16, 1945, produced such an intense blast that the sand melted and people for hundreds of miles saw the mushroom cloud. The government has cleared away the resulting radioactive material and now considers the site radiologically clean. Nearby, visitors can explore the MacDonald Ranch building where the bomb was assembled. The house has been restored to resemble its appearance when scientists worked there on the Manhattan Project that produced the bomb. Tours of the site are scheduled periodically. A visitor center and outdoor museum at Holloman Air Force Base displays missiles that have been tested here. The National Monument is also of interest for its geology and natural beauty.

MORE READING SOURCES

Allen, Frederick, et al. *American Inventions: A Chronicle of Achievements That Changed the World.* Articles reprinted from *American Heritage of Invention & Technology.* New York: American Heritage, 1995.

Amrine, Michael, et al. *Those Inventive Americans*, edited by Robert L. Breeden. Washington, D.C.: National Geographic Society, 1971.

Asimov, Isaac. *Asimov's Chronology of Science and Discovery.* New York: HarperCollins, 1994.

Billington, David P. *The Innovators: The Engineering Pioneers Who Made America Modern.* New York: John Wiley and Sons, Inc., 1996.

Cohen, I. Bernard. *Benjamin Franklin's Science.* Cambridge: Harvard University Press, 1996.

———. *Science and the Founding Fathers: Science in the Political Thought of Jefferson, Franklin, Adams, and Madison.* New York: W.W. Norton, 1995.

Gies, Joseph, and Frances. *The Ingenious Yankees.* New York: Thomas Y. Crowell Co., 1976.

Grier, Katherine, and Pat Cupples (Illustrator). *Discover: Investigate the Mysteries of History with 40 Practical Projects Probing Our Past.* Reading, Mass.: Addison-Wesley, 1990.

Jones, Charlotte Foltz, and John O'Brien (Illustrator). *Accidents May Happen: 50 Inventions Discovered by Mistake.* New York: Delacorte Press, 1996.

Machines and Inventions. (Understanding Science and Nature). New York: Time-Life, 1993.

Rennert, Richard. *Pioneers of Discovery* (Profiles of Great Black Americans). New York: Chelsea House, 1994.

INDEX

Italic page references indicate illustrations.

blast furnaces 25, 28, *31,* 33–34
Bodie Island, North Carolina 105, 113
boll weevil 58
Bond, Thomas 24
Boone, Daniel 16, 23
Boston, Massachusetts 143
botany 11, 57
British Royal Society of Literature 123
Brooke, Clement 31, 32, 34
Burbank, Elizabeth 101
Burbank, Luther 91–101, *93*
"Burbank Russet" potato 94, 99–100
Bureau of Fisheries, U.S. 122, 123
Butterfly Bush 99

C

cactus, spineless 92, 100–1, *101*
California 70, 76, 94, 98, 99, 105
California gold rush 70
California Institute of Technology 148
California State Railroad Museum 51
Campbell, W. W. 68
Canal Basin 47
cannons 27
carbohydrates 8
carbon 23
carbon dioxide 6, 8, 34
carbon monoxide 7, 34
Carnegie Institution 98
Carson, Maria McClean 119, 120, 122, 123
Carson, Marion 119, 124
Carson, Rachel 99, 117–29

Carson, Robert (brother) 119
Carson Robert Warden (father) 119, 122, 124
Carver, George Washington 53–64, 66
    statue of 56, *56,* 63–64
Carver, Jim 56, 63
Carver, Mary 55, 56
Carver, Moses 53, 54, 56, 64
Carver, Susan 53, 54, 56, 57, 64
Carver Housing Development 63
Catonsville, Maryland 144
Cayley, George 105
Chaffee, Roger 138
Chanute, Octave 107
charcoal 28, 30, *30, 31,* 33, 34
Chatham College 121
chemurgy 60
Cherokee Indians 16
cherries 99
Chesapeake and Ohio Canal National Historical Park 51–52
Chesapeake and Ohio Railroad 52
Chesapeake Bay 122
chestnut trees 96
China 134
chlorinated hydrocarbons 125
Church of England 2
Civil War 54, 56, 81
Clark, Alvan (father) 71, 73
Clark, Alvan Graham (son) 71
Cleveland, Ohio 146
climate studies 136
Clinton, George 3
Cloisters, The 102
Colorado 70

Columbia, Pennsylvania 42
Columbia River 147
Comet Mitchell 1847 VI 76
Congress, U.S. 34, 62, 64, 145
Constitution Square State Historic Site 23–24
Continental Army 27
cotton 58, 60
Coulee Dam, Washington 147
Coulee Dam National Recreation Area 147
Craddock, Robert 24
Crawford, Jane Todd 19–21
Creek Indians 16
Cresson, Pennsylvania 37, 51
Cresson Summit 43
crop rotation 60
cross-breeding 94
Cumberland, Maryland 52
Cumberland Gap 16
Curie, Marie 88
Cyrus McCormick Farm and Workshop 146

D

Danville, Kentucky 14, 16, 18, 20, 23–24
Darwin, Charles 92, 94
Darwin, Erasmus 11
Davidson, George 70
Davy, Humphrey 7, 83
Dayton, Ohio 105, 106, 108, 112, 113, 115, 116
Dayton Aviation Heritage National Historic Park 116
DDT 124–25
Dearborn, Michigan 77, 79, 81, 86, 116, 147
de Camp, L. Sprague 79

W

Washington, Booker T. 58
Washington, George 27
Washington, D.C. 52, 83, 122, 139, 144, 146
Washington State 147
Watkins, Andrew 57
Watkins, Mariah 57
Weightless Environment Training Facility (WET-F) 141
Wells, Maine 130
Western Union 81, 82
Westinghouse, George 85–86
West Orange, New Jersey 85, 89
White, Canvass 40
White, Edward 138

White Sands Missile Range Trinity Site 148
White Sands National Monument 148–49
Wilderness Road 16, 23
Williamsport, Maryland 52
Willis Russell House 24
Woman's Home Companion magazine 124
Woods Hole, Massachusetts 121
Worcester, Massachusetts 94
World War I 62, 112
World War II 122–23, 134
Wright, Orville 103–114, 115, 116, 147
Wright, Wilbur 103–114, 115, 116, 147

Wright Brothers National Memorial 103–15, *110, 111, 113*
Wright Flyer III 116
Wright-Patterson Air Force Base 116
wrought iron 34
Wyoming 70, 83

Y

Yale Review 123
Yerkes Observatory 73
Yosemite National Park 69
Young, John 133

Z

zoology 11, 121